AGRICULTURE ISSUES AND POLICIES

SPECIALTY CROPS

FEDERAL PROGRAMS AND INSURANCE

AGRICULTURE ISSUES AND POLICIES

Additional books in this series can be found on Nova's website under the Series tab.

Additional E-books in this series can be found on Nova's website under the E-book tab.

AGRICULTURE ISSUES AND POLICIES

SPECIALTY CROPS

FEDERAL PROGRAMS AND INSURANCE

KATELIN FRANSEN
EDITOR

nova publishers
New York

For permission to use material from this book please contact us:
Telephone 631-231-7269; Fax 631-231-8175
Web Site: http://www.novapublishers.com

Library of Congress Cataloging-in-Publication Data

ISBN: 978-1-62618-054-3

Published by Nova Science Publishers, Inc. † New York

Contents

Preface		**vii**
Chapter 1	Fruits, Vegetables, and Other Specialty Crops: Selected Federal Programs *Renée Johnson*	**1**
Chapter 2	Federal Crop Insurance for Specialty Crops: Background and Legislative Proposals *Dennis A. Shields*	**67**
Chapter 3	Report to Congress: Specialty Crop Report *United States Department of Agriculture*	**93**
Index		**135**

PREFACE

U.S. farmers grow more than 350 types of fruit, vegetable, tree nut, flower, nursery, and other horticultural crops in addition to the major bulk commodity crops. Specialty crop producers are ineligible for the federal commodity price and income support programs that benefit commodity crop producers (e.g. grains and cotton); however, they are eligible for other types of USDA support. Unlike federal support for commodity crops, support for specialty crops spans a wide range of existing USDA programs, many of which also provide support to other agricultural commodities. These include marketing and promotion programs, crop insurance and disaster assistance, plant pest and disease protections, trade assistance, and research and extension services. This book examines specialty crop federal programs and insurance with a focus on their background and legislative proposals.

Chapter 1 - U.S. farmers grow more than 350 types of fruit, vegetable, tree nut, flower, nursery, and other horticultural crops in addition to the major bulk commodity crops. Specialty crop producers are ineligible for the federal commodity price and income support programs that benefit commodity crop producers (e.g., grains and cotton); however, they are eligible for other types of U.S. Department of Agriculture (USDA) support. Unlike federal support for commodity crops, support for specialty crops spans a wide range of existing USDA programs, many of which also provide support to other agricultural commodities. These include marketing and promotion programs, crop insurance and disaster assistance, plant pest and disease protections, trade assistance, and research and extension services, among other types of miscellaneous support. The industry also benefits from fruit and vegetable purchases under various domestic nutrition assistance programs. Despite this wide range of program support, overall program spending on all specialty crops remains a small fraction of that spent on all commodity crops, even when considering both mandatory and discretionary funding.

Some of the programs supporting specialty crops are longstanding farm support programs that benefit all agricultural producers and are regularly contained within omnibus farm legislation. However, several programs addressing specialty crops specifically were established following the enactment of the Specialty Crops Competitiveness Act of 2004 (P.L. 108-465), which was enacted outside a farm bill year. Many of the programs in the 2004 act were further expanded and reauthorized in the 2008 farm bill (Food, Conservation, and Energy Act of 2008, P.L. 110-246). Other programs were established in the 2002 farm bill (Farm Security and Rural Investment Act of 2002, P.L. 107-171), often as pilot initiatives that have since become established programs. Other laws, such as the Perishable Agricultural Commodities Act of 1930 (PACA) and the Agricultural Marketing Agreement Act of 1937, were enacted long ago to exclusively serve the produce industry to protect sellers in the marketplace. Other federal agencies also play important roles in the specialty crop industry. The Food and Drug Administration (FDA, in the U.S. Department of Health and Human Services) is responsible for assuring that fresh, frozen, canned, and imported fruits, vegetables, and nuts are safe for human consumption. Recently enacted food safety reforms (FDA Food Safety Modernization Act, FSMA) placed additional regulatory requirements on certain specialty crop growers and processors to comply with safety requirements for foods that are regulated by FDA, which includes specialty crops. Under FSMA, FDA is developing mandatory food safety regulations and traceability requirements affecting farmers, packers, and processors of both domestically produced and imported foods under FDA's jurisdiction. At the farm production level, these requirements will mostly affect produce growers.

Among other agencies, the Environmental Protection Agency sets the safe limits for pesticide residues on produce, which FDA enforces. The Department of Commerce and the International Trade Commission are responsible for investigating instances of suspected "dumping" of foreign goods on the U.S. market and levying antidumping taxes. The Department of Labor, the Department of Homeland Security, and the Department of State jointly administer a system for temporarily admitting foreign workers to provide seasonal labor, provided that U.S. workers are not available.

Chapter 2 - The federal crop insurance program provides farmers with risk management tools to address crop yield and/or revenue losses on their farms.

Farmers can purchase subsidized policies that pay an indemnity when their production or revenue falls below a guaranteed level.

Historically, the federal crop insurance program primarily has covered traditional farm program crops such as wheat, corn, and soybeans. However, the crop insurance program has expanded in recent decades, and policies are available now for a wide range of commodities, including specialty crops such as fruits and vegetables. In 2009, specialty crop policies covered more than 7 million acres, which was roughly one-half to three-quarters of specialty crop area, depending on how total area is calculated. In total, crop insurance is available for over 80 specialty crops, making the program the primary financial safety net for specialty crop producers.

The federal crop insurance program is permanently authorized by the Federal Crop Insurance Act, as amended (7 U.S.C. 1501 et seq.). The U.S. Department of Agriculture's (USDA's) Risk Management Agency (RMA) operates the Federal Crop Insurance Corporation (FCIC), which is the funding mechanism for the program. Insurance policies are sold and completely serviced through approved private insurance companies. The insurance companies' losses are reinsured by USDA, and their administrative and operating costs are reimbursed by the federal government.

Federal crop insurance policies for specialty crops (and other crops as well) are generally either yield-based or revenue-based. For most yield-based policies, a producer can receive an indemnity if there is a yield loss relative to the farmer's "normal" (historical) yield. Insurable causes of loss include drought, excess precipitation, hail, frost, freeze, fire (if due to natural causes), and insects and disease. Revenue-based policies protect against crop revenue loss resulting from declines in yield, price, or both. Nursery crop producers can be protected against loss of an "asset" due to weather damage (but not price loss). Also relevant for specialty crop producers is whole-farm insurance, which protects against losses in whole farm revenue rather than just for an individual crop. An endorsement is available in some areas for protecting against loss due to quarantine.

While additional policies have been introduced in the last 10 years, producers and some Members of Congress would like to enhance crop insurance for specialty crop producers. In the 112th Congress, several legislative proposals have been introduced. The Local Farms, Food, and Jobs Act of 2011 (H.R. 3286/S. 1773) would enhance whole farm insurance and insurance for organic crops. The Rural Economic Farm and Ranch Sustainability and Hunger (REFRESH) Act of 2011 (S. 1658 and H.R. 3111) would also enhance whole farm insurance. The Specialty Crop Insurance Act of 2011 (S. 1905) would support the development of new policies.

In the 112th Congress, specialty crop provisions in these and other proposals have been included in the Senate-passed farm bill (S. 3240) and the House committee-passed farm bill (H.R. 6083). The bills require USDA to conduct more research on higher coverage levels for whole farm revenue insurance and study insurance for food safety and contamination-related losses. A provision would revise the value of crop insurance for all organic crops to reflect prices of organic (not conventional) crops. For crops not insurable with crop insurance, additional coverage would become available under the existing Noninsured Crop Assistance Program (NAP). Finally, a provision in S. 3240 would subsidize the purchase of existing private-sector index-based weather insurance, which protects against specific weather events and not actual loss.

Chapter 3 - The 2009 Specialty Crop Report is submitted in response to Section 508 (a)(6)(B) of the Federal Crop Insurance Act (Act), as amended, requiring that the Federal Crop Insurance Corporation (FCIC) report to Congress on the progress and expected timetable for expanding crop insurance coverage under the Act to new and specialty crops. As indicated by the requirements of the law, the report is a useful way to obtain a quick overview of the processes and timelines the Risk Management Agency (RMA) must follow in order to make new and specialty crop insurance products available to producers.

This report complements *Participation in the Federal Crop Insurance Program* (Participation Report) submitted in accordance with Section 508(a)(7) of the Act. The Participation Report was prepared to assist the FCIC Board of Directors (Board) in establishing development priorities for the future. The Participation Report also outlined the progress that the FCIC has made since a comprehensive analysis was completed of the crop insurance market. As detailed in the Participation Report, the Board questioned whether FCIC may be nearing a saturation point where the potential for cost effectively adding new crop programs is limited. However, as new industries emerge as legitimate agricultural enterprises, RMA continually monitors the need for additional coverage options for these developing industries.

The growth of specialty crop programs can be measured by the frequency with which new and/or revised product offerings for specialty crops are considered by the Board. During calendar year 2009 alone, the Board was presented with new or modified products for the following specialty commodities: apiculture, avocados, citrus, dry peas & lentils, fresh market beans, strawberries, sugar beets, sugarcane, and tobacco.

Each year RMA responds to a number of commodity organizations and industry representatives who either desire a new Federal crop insurance product or an enhancement to the existing coverage options. Enhancement requests often focus on revenue coverage, which was first made available to the major commodities (e.g., corn, soybeans) beginning in the mid-1990's and bases coverage on values obtained from commodity futures exchanges. These major commodities have since experienced a rapid shift from yield based to revenue coverage over the past decade which is evidence of the demand for revenue coverage among agricultural producers. However, there is no such comparable price discovery mechanism for specialty crops, and therefore development of a viable revenue insurance product is much more difficult. Nevertheless, RMA remains committed to exploring all options for possibly providing revenue coverage to specialty crops. For example, the Board recently approved a new revenue coverage product, Actual Revenue History (ARH), for cherries, navel oranges and strawberries. Given its design, the ARH product is most appropriate for fresh fruit and perishable commodities. Conversely, the applicability of ARH beyond this niche can be problematic because basic assumptions of the ARH design are violated.

In considering new crop programs designed to benefit specialty crop producers, RMA and the Board are faced with the challenge of allocating limited resources across competing interests. This task has become more challenging given the requirements of Office of Management and Budget Memorandum, "Budget Discipline for Agency Administrative Actions", dated May 23, 2005, commonly known as PAYGO requirements, and RMA's relatively static operational budget. Further, expanding existing specialty crop programs to additional growing areas also requires meeting PAYGO requirements because these expansions are discretionary actions. Thus, without a corresponding reduction, or savings, in another part of the crop insurance program or in another USDA program, new and/or expanded programs cannot be implemented because funding is not available.

Low participation, high loss ratios, and program design limitations learned through the pilot program process have resulted in the Board terminating several crop insurance programs. Program terminations over the past several years include: crambe, processing cucumbers, fresh market beans, raspberries, blackberries, sweet potatoes, and winter squash.

In: Specialty Crops
Editor: Katelin Fransen
ISBN: 978-1-62618-054-3

Chapter 1

FRUITS, VEGETABLES, AND OTHER SPECIALTY CROPS: SELECTED FEDERAL PROGRAMS*

Renée Johnson

SUMMARY

U.S. farmers grow more than 350 types of fruit, vegetable, tree nut, flower, nursery, and other horticultural crops in addition to the major bulk commodity crops. Specialty crop producers are ineligible for the federal commodity price and income support programs that benefit commodity crop producers (e.g., grains and cotton); however, they are eligible for other types of U.S. Department of Agriculture (USDA) support. Unlike federal support for commodity crops, support for specialty crops spans a wide range of existing USDA programs, many of which also provide support to other agricultural commodities. These include marketing and promotion programs, crop insurance and disaster assistance, plant pest and disease protections, trade assistance, and research and extension services, among other types of miscellaneous support. The industry also benefits from fruit and vegetable purchases under various domestic nutrition assistance programs. Despite this wide range of program support, overall program spending on all specialty crops remains a small

* This is an edited, reformatted and augmented version of a Congressional Research Service publication, CRS Report for Congress R42771, prepared for Members and Committees of Congress, from www.crs.gov, dated October 3, 2012.

fraction of that spent on all commodity crops, even when considering both mandatory and discretionary funding.

Some of the programs supporting specialty crops are longstanding farm support programs that benefit all agricultural producers and are regularly contained within omnibus farm legislation. However, several programs addressing specialty crops specifically were established following the enactment of the Specialty Crops Competitiveness Act of 2004 (P.L. 108-465), which was enacted outside a farm bill year. Many of the programs in the 2004 act were further expanded and reauthorized in the 2008 farm bill (Food, Conservation, and Energy Act of 2008, P.L. 110-246). Other programs were established in the 2002 farm bill (Farm Security and Rural Investment Act of 2002, P.L. 107-171), often as pilot initiatives that have since become established programs. Other laws, such as the Perishable Agricultural Commodities Act of 1930 (PACA) and the Agricultural Marketing Agreement Act of 1937, were enacted long ago to exclusively serve the produce industry to protect sellers in the marketplace. Other federal agencies also play important roles in the specialty crop industry. The Food and Drug Administration (FDA, in the U.S. Department of Health and Human Services) is responsible for assuring that fresh, frozen, canned, and imported fruits, vegetables, and nuts are safe for human consumption. Recently enacted food safety reforms (FDA Food Safety Modernization Act, FSMA) placed additional regulatory requirements on certain specialty crop growers and processors to comply with safety requirements for foods that are regulated by FDA, which includes specialty crops. Under FSMA, FDA is developing mandatory food safety regulations and traceability requirements affecting farmers, packers, and processors of both domestically produced and imported foods under FDA's jurisdiction. At the farm production level, these requirements will mostly affect produce growers.

Among other agencies, the Environmental Protection Agency sets the safe limits for pesticide residues on produce, which FDA enforces. The Department of Commerce and the International Trade Commission are responsible for investigating instances of suspected "dumping" of foreign goods on the U.S. market and levying antidumping taxes. The Department of Labor, the Department of Homeland Security, and the Department of State jointly administer a system for temporarily admitting foreign workers to provide seasonal labor, provided that U.S. workers are not available.

INTRODUCTION

Specialty crops, defined as "fruits and vegetables, tree nuts, dried fruits, and horticulture and nursery crops (including floriculture),"[1] comprise a major part of U.S. agriculture. In 2007, the value of farm-level specialty crop production totaled more than $50 billion, representing more than one-third of the value of U.S. crop production (Table 1). Despite their relatively large share of crop receipts, specialty crops occupy only about 3% of U.S. harvested cropland acres.[2]

The U.S. Department of Agriculture (USDA) reports that retail sales of fresh and processed fruits and vegetables for at-home consumption total nearly $100 billion annually.[3]

Exports of U.S. specialty crops totaled nearly $15.9 billion in 2010, or about 15% of total U.S. agricultural exports.[4]

In 2007, about 248,000 farming operations grew more than 350 types of fruit, vegetable, tree nut, flower, nursery, and other horticultural crops in addition to the major bulk commodity crops.[5] Farm sales are focused in California, Florida, Washington, Oregon, North Dakota and Michigan; however, every state has some commercial specialty crop production within its borders. USDA data illustrate the distribution, nationwide, of areas producing vegetables (Figure 1), fruits and tree nuts (Figure 2), and nursery crops (Figure 3), shown as a percentage of the total market value of agricultural products sold (including livestock).

The majority of specialty crop producers are considered specialized, which means that they receive at least half of their gross value of production from the sale of vegetables, fruits, tree nuts or other horticultural crops. These specialized farms rely mostly on specialty crop production for their farm income, even though they may be also engaged in other forms of agricultural production. USDA reports that about 50% of all vegetable growers and 80% of fruit and tree nut growers are considered specialized; however, specialized farms account for 90%-95% of the total value of U.S. specialty crop production.[6]

Conditions may vary considerably by major production region. Specialized fruit and vegetable farms are more concentrated in the western United States, including California, Washington, and Oregon. Some farms also participate in the major commodity support programs, but these tend to be more concentrated in the midwestern states.

Even though specialty crop production accounts for about one-third of the value of all U.S. crop production, overall spending on federal programs benefitting all specialty crops remains a small fraction of that spent on all commodity crops.[7]

Table 1. U.S. Crop Production Statistics, Commodity and Horticultural Crops

		2007	
	Farms (1,000)	Sales ($ billion)	Sales (%)
Total U.S. Agriculture			
Crops, including nursery and greenhouse	986	143.7	48%
Livestock, poultry, and their products	1,080	153.6	52%
Total	2,205	297.2	100%
Commodity crops			
Grains, oilseeds, dry beans, dry beans	479	77.2	54%
Cotton & Tobacco	35	6.2	4%
Other crops and hay	435	10.0	7%
Subtotal	949	93.3	65%
Specialty Crops			
Vegetables, melons, potatoes	69	14.7	10%
Fruits, tree nuts, and berries	113	18.6	13%
Nursery, greenhouse, floriculture	51	16.6	12%
Cut trees, and short rotation woody crops	13	0.4	0%
Subtotal	246	50.3	35%
Total, Crops, incl. nursery, greenhouse	986	143.7	100%

Source: USDA, NASS, 2007 Census of Agriculture, Specialty Crops, Volume 2, November 2009. Table 2, http://www.agcensus.usda.gov/Publications/2007/Full_Report/Volume_1,_Chapter_1_US/st99_1_002_002.pdf

Note: The total number of farms does not add since the totals include other types of farming operations.

Precise estimates of total mandatory and discretionary sources of funding are difficult to measure, given that support for specialty crops is spread across a wide range of USDA programs and not within a price and income support program such as that available for most of the major commodity crops. Following the 2008 farm bill, an average of approximately $0.676 billion annually (FY2008-FY2012) in mandatory program funding was authorized to be spent on specialty crops, mostly through government purchases of fruits and vegetables for domestic nutrition and feeding programs (*Table 2*).

(% of total market value of agricultural products sold, 2007)

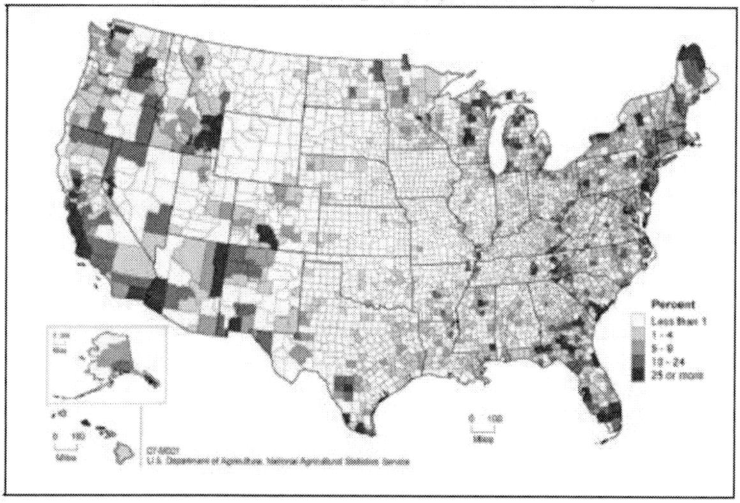

Source: USDA, NASS, 2007 Census of Agriculture, Specialty Crops, vol. 2, November 2009.

Figure 1. Value of Vegetables, Melons, Potatoes, and Sweet Potatoes Sold (% of total market value of agricultural products sold, 2007).

(% of total market value of agricultural products sold, 2007)

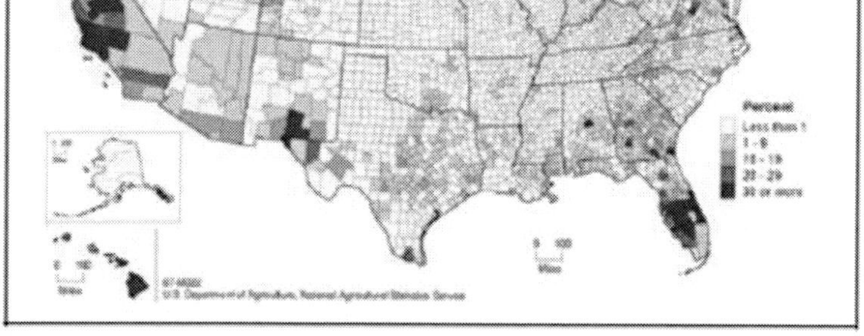

Source: USDA, NASS, 2007 Census of Agriculture, Specialty Crops, vol. 2, November 2009.

Figure 2. Value of Fruits, Tree Nuts, and Berries Sold.

This compares to estimates for the major commodity crops, which received approximately $8.3 billion annually mostly through direct price and income support.[8]

The specialty crop industry is requesting that Congress provide additional funding for certain existing programs, reaching approximately $0.8 billion annually.[9]

This represents an increase above current funding levels for specialty crops, but would still represent a very small share of estimated total farm bill spending compared to commodity crops.

(% of total market value of agricultural products sold, 2007)

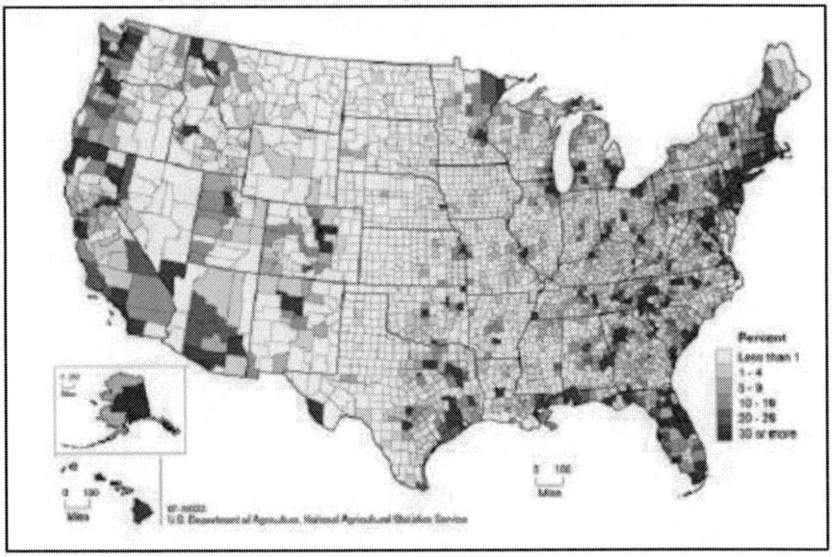

Source: USDA, NASS, 2007 Census of Agriculture, Specialty Crops, vol. 2, November 2009.

Figure 3. Value of Nursery, Greenhouse, Floriculture and Sod Sales.

Selected Federal Programs

Specialty crops are ineligible for the federal commodity price and income support programs that benefit producers of commodity crops; however, they are eligible for other types of USDA support. Unlike programs intended to support the specific commodity crops, programs supporting specialty crops are generally available to all crops.

Table 2. Selected Authorized Funding Levels for Specialty Crops and Organic Agriculture, FY2008-FY2012 ($ millions)

Program Name (2008 farm bill section number)	Funding Type[a]	Average Annual (Mand.)	FY2008-FY2012 (Mand.)
Specialty Crops			
Technical Assistance for Specialty Crops, TASC (Sec. 3203)	Mandatory	$7.4	$37.0
Fresh Fruit and Vegetable ("Snack") Program (Sec. 4304)	Mandatory/Discretionaryb	$101.2	$506.0
Minimum Purchases under Section 32 program (Sec. 4404)	Mandatory	$398.2	$1,991.0
Senior Farmers' Market Program (Sec. 4231)	Mandatory	$20.6	$103.0
SNAP Pilot Projects (Sec. 4141)	Mandatory/Discretionary[b]	$4.0	$20.0
Value-Added Producer Grants, VAPG (Sec. 6202)	Mandatory/Discretionaryc	$3.0	$15.0
Specialty Crop Research Initiative, SCRI (Sec. 7311)	Mandatory/Discretionaryc	$46.0	$230.0
Specialty Crop Block Grants (Sec. 10109)	Mandatory	$44.8	$224.0
Farmers' Market Promotion Program (Sec. 10106)	Mandatory	$6.6	$33.0
Specialty Crops Market News Allocations (Sec. 10107)	Mandatory	$9.0	$45.0
Food Safety Education Initiatives (Sec. 10105)	Discretionary[d]	na	na
Plant Pest/Disease Mgmt./Disaster Prevention (Sec. 10201)	Mandatory	$31.4	$157.0
National Clean Plant Network (Sec. 10202)	Mandatory	$4.0	$20.0
Subtotal		$676.2	$3,381.0
Certified Organic (all crops and livestock)[e]			
Organic Agric. Research and Extension Initiative (Sec. 7206)	Mandatory/Discretionary[b]	$15.6	$78.0
Nat'l Organic Certification Cost-Share Program (Sec. 10301)	Mandatory	$4.4	$22.0
Organic Prodt./Market Data Initiative (ODI) (Sec. 10302)	Mandatory/Discretionary	$1.0	$5.0
National Organic Program, NOP (Sec. 10303)	Discretionary	na	na
Subtotal		$21.0	$105.0
TOTAL		$697.2	$3,486.0

Source: CRS, from the 2008 farm bill (P.L. 110-246). Section numbers shown in parentheses. Average annual is the simple average over the five year period. Data for FY2008-FY2012 are totals, and include program funding that may have been allocated on a one-time basis (often made available until expended). Excludes programs that are available to all agricultural producers (such as conservation programs, trade promotion programs, etc.) where the specialty crop portion is not readily identifiable.

Notes: Data in the table reflect mandatory funding only and not authorized appropriations for some programs.

[a] Mandatory funding is made available by multiyear authorizing legislation and does not require annual appropriations or subsequent action by Congress. Discretionary spending requires appropriations action.

[b] Authorized appropriations of "such sums as necessary."

[c] Authorized annual appropriations (FY2008-FY2012) of $40 million (VAGP), $100 million (SCRI), and $5 million (ODI). NOP appropriations authorized at $5 million (FY2008) rising to $11 million (FY2012).

[d] Authorizes appropriations of $1 million/year (FY2008–FY2012), "to remain available until expended".

[e] Spending for certified organic agriculture includes funding for all certified organic production, which includes meat and dairy foods, as well as organic commodity crops, in addition to fruits and vegetables.

These include marketing and promotion programs, crop insurance and disaster assistance, plant pest and disease protections, trade assistance, and research and extension services, among other types of miscellaneous support. The industry also benefits from fruit and vegetable purchases under various food and nutrition programs.

Some of the programs supporting specialty crops are longstanding farm support programs that benefit all agricultural producers and are regularly contained within omnibus farm bill legislation. However, several programs specifically addressing specialty crops were established following the enactment of the Specialty Crops Competitiveness Act of 2004 (P.L. 108-465), which was enacted outside a farm bill year.

Many of the programs in the 2004 act were further expanded and amended in the 2008 farm bill.[10] Other programs were established in the 2002 farm bill[11]—many of which were started as pilot initiatives that have since become established programs.

Other laws, such as the Perishable Agricultural Commodities Act of 1930 (PACA), were enacted long ago to exclusively serve the produce industry to protect sellers in the marketplace.

USDA programs supporting specialty crop producers are spread across many different titles of the farm bill (see text box above). Many of these programs broadly apply to a range of agricultural commodities, including fruits, vegetables, and other specialty crops.

Selected Specialty Crop Provisions in the 2008 Farm Bill (P.L. 110-246)

Commodities (Title I)

- Planting Flexibility (§§ 1107, 1306)

Conservation (Title II)

- Adjusted Gross Income (AGI) limit (Conservation) (§ 1604)
- Conservation programs incentives (§§ 2701, 2509, 2706, 2707)

Trade (Title III)

- Market Access Program (MAP) (§ 3102)
- Technical Assistance for Specialty Crops (TASC) (§ 3203)

Nutrition (Title IV)

- Fresh Fruit and Vegetable ("Snack") Program (§ 4304)
- Minimum Purchases under Section 32 program (§ 4404)
- Senior Farmers' Market Nutrition Program (§ 4231)
- Purchases of Locally Produced Foods (§ 4302)
- Supplemental Nutrition Assistance Program (SNAP) pilot (§ 4141)

Rural Development (Title VII)

- Value-Added Producer Grants (§ 6202)

Research (Title VII)

- Specialty Crop Committee Report (§ 7103)
- Specialty Crop Research Initiative (§ 7311)
- Office of Pest Management Policy (§ 7313)

Horticulture and Organic Agriculture (Title X)

- Specialty Crop Block Grants (§ 10109)
- Farmer's Market Promotion Program (§ 10106)
- Specialty Crops Market News Allocations (§ 10107)
- Evaluation of USDA Commodity Purchase Process (§ 10101)
- Inclusion of Specialty Crops in Census of Agriculture (§ 10103)
- Food Safety Education Initiatives (§ 10105)
- Plant Pest and Disease Mgmt. and Disaster Prevention (§ 10201)
- National Clean Plant Network (§ 10202)
- Plant Protection Amendment (§ 10203)
- Invasive Pest and Disease Emergency Response Funding (§ 10206)
- Grants to Improve the Movement of Specialty Crops (§ 10403)

Livestock (Title XI)

- Country of Origin Labeling (COOL) (§ 11002)

Crop Insurance and Disaster Assistance Programs (Title XII)

- Tree Assistance Program (TAP); and Emergency Assistance for Livestock, Honeybees, and Farm-Raised Fish Program (§§ 12033, 15101)

Source: Compiled by CRS.

The selected programs described in this report are administered by various USDA agencies, including the Agricultural Marketing Service (AMS), Animal and Plant Health Inspection Service (APHIS), Food and Nutrition Service (FNS), Risk Management Agency (RMA), Farm Service Agency (FSA), National Institute of Food and Agriculture (NIFA), Agricultural Research Service (ARS), Natural Resources Conservation Service (NRCS), Foreign Agricultural Service (FAS), and Rural Development (RD).

Specialty crop producers likely also benefit from other USDA programs, available to all agricultural producers, that are not specifically highlighted in this report. These include other USDA research and cooperative extension programs, as well as USDA conservation and rural development programs, among others.

Other federal agencies play a role in the specialty crop industry. These include agencies that oversee food safety requirements for fruits, vegetables, and other specialty crops, such as the Food and Drug Administration and the Environmental Protection Agency, and agencies that oversee global trade, such as the Department of Commerce and the U.S. International Trade Commission, among others.

Following is a description of the key USDA programs, as well as programs administered by other federal agencies. Where applicable, a primary source of information on these selected programs is the Catalog of Federal Domestic Assistance.[12]

Advisory Committee

USDA established a Fruit and Vegetable Industry Advisory Committee in August 2001, which is currently re-chartered through 2013. The purpose of the committee is to examine the full spectrum of issues faced by the industry and to provide suggestions on how USDA can tailor its programs to better meet the industry's needs.[13] The committee holds open meetings, which AMS announces in advance in the *Federal Register*. Up to 25 members may be appointed, consisting of those who represent the fruit and vegetable industry, including fruit and vegetable growers/shippers, wholesalers, brokers, retailers, processors, fresh cut processors, food-service suppliers, state agencies involved in organic and non-organic fresh fruits and vegetables at local, regional, and national levels, state departments of agriculture, and trade associations. Committee members are appointed by USDA and serve two- to three-year terms.

Assistance for Production Losses

Federal Crop Insurance

USDA's Risk Management Agency (RMA) administers the federal crop insurance program.[14] Approved private insurance companies sell and completely service the policies, but USDA reinsures potential losses and either fully or partially compensates the companies for any losses incurred. Eligible producers can receive catastrophic insurance, which is basically free except for an administrative fee. Producers can buy up their level of coverage beyond the catastrophic level and pay a premium that is subsidized by the federal government. Revenue insurance, which makes indemnity payments for income lost either from poor production or low market prices, also is available to producers of certain crops in many areas. Such insurance provides an indemnity payment when actual revenue falls below a target level of revenue.[15] USDA decides which crops in which geographical areas will be covered by which types of insurance. The decision is made on a crop-by-crop and county-by-county basis, based on farmer demand for coverage and the level of risk associated with the crop in the region, among other factors. The RMA frequently offers pilot programs with various types of coverage for new crops (particularly specialty crops) or new geographical areas. It uses the performance of these programs to inform its decision on whether to extend coverage permanently.

USDA estimates that, as of 2009, 7.25 million acres of fruits, vegetables, and tree nuts were covered by federal crop insurance, with about 75% share of acres insured across all covered specialty crops (*Figure 4*). This estimate is based on total acreage for examined specialty crops only.[16] If the number of insured acres were matched against the total number of production acres (i.e., regardless of crop insurance availability), this estimate would be much lower at about 58%.[17] This compares with an estimated 83% coverage for major commodity crops (i.e., those under farm commodity price and income support programs). In addition, actual coverage varies depending on crop type. The share of acres insured is greater than 90% for some specialty crops (raisins, prunes, and citrus fruit), but lower than 50% for other crops (cabbage, peppers, processing beans, pumpkins, tropical fruit, and walnuts; see *Figure 4*).

Roughly, about 80 types of fruits, vegetables, and tree nuts are currently covered by individual federal crop insurance plans.[18] Crops covered by individual federal crop insurance plans include almonds, apples, apricots, avocados, bananas, certain beans, blueberries, citrus, cherries, cranberries, figs, grapes (including raisins), macadamia nuts, nectarines, onions, papayas, peaches, pecans, peppers, plums (including prunes), potatoes, pumpkins, sweet corn, certain tomatoes, and walnuts. Other specialty crops, estimated at more than 1 million acres, do not have insurance available.[19] These include asparagus, beets, most types of berries (including strawberries), broccoli, carrots (fresh and for processing), cashews, cauliflower, celery, chives, cucumbers (fresh and for processing), dates, eggplants, garlic, hazelnuts, leeks, lettuce, melons, olives, pistachios, spinach and other leafy greens, squash, tropical plants, and most other root plants.20 Some of these crops may be covered by other types of insurance coverage, such as plans based on historical farm income (e.g., whole farm insurance programs).

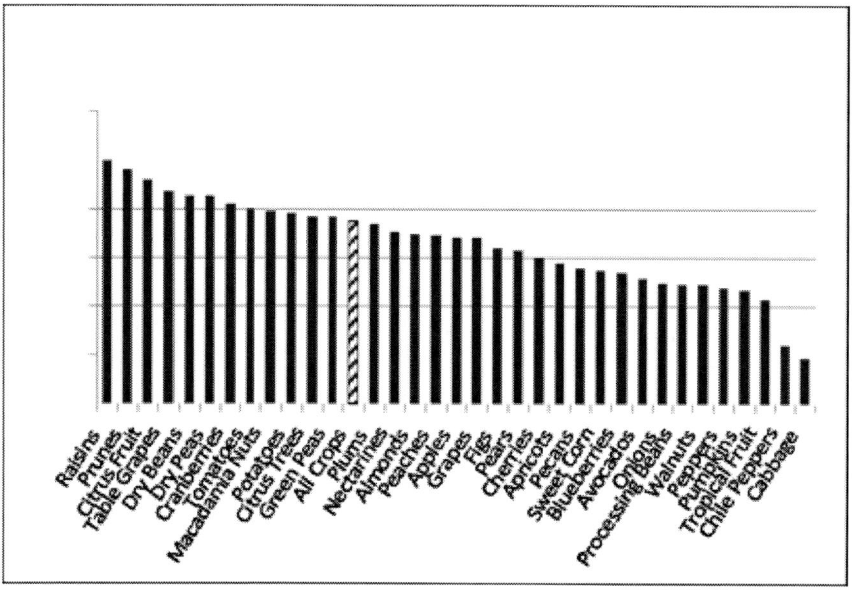

Source: Data are from USDA, FCIC, Report to Congress: Specialty Crop Report, November 2010. Graphic is from Keith Collins, National Crop Insurance Services, "Crop Insurance for Specialty Crops." Presentation to the Specialty Crop Caucus, House of Representatives, April 12, 2012.

Figure 4. Share of Specialty Crop Acres Insured, 2009.

Participation among specialty crop producers is relatively high in major producing states, including California (71% of total crop area), Florida (91%), and Washington (68%). Participation for pulse crops (e.g., dry peas, dry beans) is high in the major growing states in the Northern Plains, including Minnesota (84%), Montana (83%), North Dakota (96%), and South Dakota (79%). Other states with specialty crop production (and their participation rates) include Michigan (73%), New York (70%), and Oregon (52%). These figures compare with 83% participation for major program crops (i.e., those under farm commodity price and income support programs).[21]

In FY2011, premium subsidies received by all U.S. agricultural producers totaled $7.5 billion. Of this total, specialty crops received an estimated $425 million.[22]

Reportedly, although fruits and vegetables (not including nursery crops) account for 22% of estimated 2011 U.S. crop receipts, fruits and vegetables account for less than 5% of annual crop insurance premiums.[23] This compares to 87% of crop insurance premiums for selected commodity crops (corn, soybeans, wheat, and cotton), which together account for about 58% of annual U.S. crop receipts.

Many of the new crop insurance products introduced each year are intended to broaden coverage of fruits, vegetables, and tree nuts (*Figure 5*). Specialty crop growers reportedly face a number of challenges pertaining to expanding insurance coverage, including generally small acreages (a marketability issue compared to that for commodity crops); multiple crop varieties and farming practices (which contribute to greater complexity and cost); quality and price discovery issues; concerns about grower interest; non-weather risks; and other coverage limitations.[24]

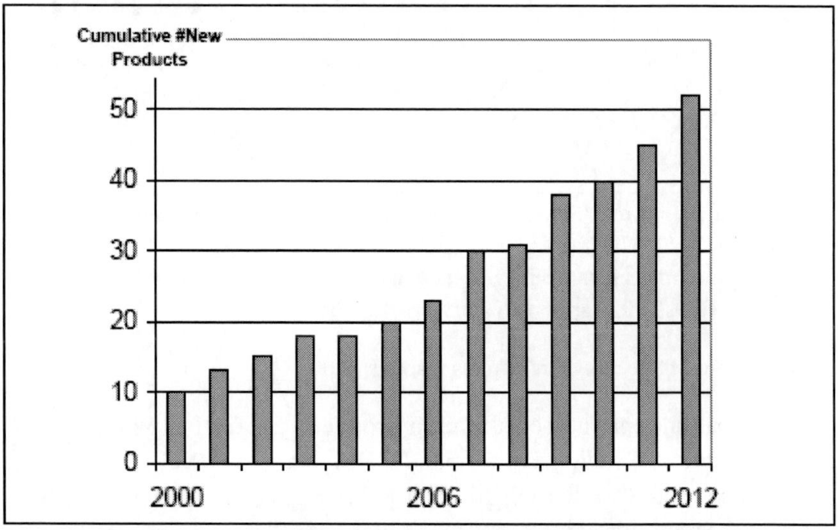

Source: Keith Collins, National Crop Insurance Services, "Crop Insurance for Specialty Crops." Presentation to the Specialty Crop Caucus, House of Representatives, April 12, 2012. Figure shows cumulative number of new products each year (2000-2012). In 2012, new products added included popcorn, strawberries, tangerine trees Citrus VI, camelina, pistachios, olives.

Figure 5. New Crop Insurance Product Introductions, by Year.

Noninsured Disaster Assistance

Producers of any commercial crops that are not insurable under the federal crop insurance program are potentially eligible for payments up to $100,000 per person under USDA's noninsured assistance program (NAP).[25] USDA's Farm Service Agency (FSA) administers NAP, which has permanent authority under the Federal Crop Insurance Reform Act of 1994 (P.L. 103- 354, as amended).[26] Specialty crops currently eligible for the NAP include mushrooms, flowers, ornamental nursery crops, Christmas trees, turfgrass sod, and ginseng, among other specialty crops. An individual producer is ineligible if the farmer's average nonfarm adjusted gross income exceeds $500,000. NAP is not subject to annual appropriations, but rather is a mandatory program that receives such sums as necessary through USDA's Commodity Credit Corporation (CCC).[27]

In FY2011, USDA estimates it made $71 million in NAP payments to all U.S. agricultural producers.[28] Breakouts by individual commodities or commodity groupings are not available.

Other Supplemental Assistance

The 2008 farm bill authorized three supplemental agricultural disaster assistance programs administered by FSA that provide assistance to specialty crop growers for certain losses that occurred on or after January 1, 2008, and before October 1, 2011.[29]

Each of these three programs expired in 2011 and it remains unclear whether they will be reauthorized.

- The Tree Assistance Program (TAP) provides financial assistance to qualifying orchardists and nursery tree growers to replant or rehabilitate eligible trees, bushes, and vines damaged by natural disasters, if mortality losses are in excess of 15% (after adjustment for normal mortality). It increases the maximum payment to qualifying orchardists and nursery growers to $100,000 a year to reflect tree removal and replacement costs.[30]
- The Emergency Assistance for Livestock, Honeybees, and Farm-Raised Fish Program (ELAP) provides assistance to beekeepers who might provide pollination services for specialty crop growers.[31] Coverage includes losses from disaster such as adverse weather or other conditions (such as blizzards and wildfires) that are not adequately covered by any other disaster program.
- The Supplemental Revenue Assistance Payments (SURE) program provides supplemental revenue assistance for crop producers to compensate eligible producers for a portion of crop losses that are not eligible for an indemnity payment under the crop insurance program (i.e., the portion of losses that is part of the deductible on the policy).[32] Specialty crop producers were also included in the program's revenue guarantee and loss calculations.

FSA also administers a program that makes low-interest emergency loans to farmers in counties that have been officially declared disaster areas.[33] FSA provides such loans to help producers recover from production losses or physical losses. In the case of specialty crops, destruction of established fruit trees, or buildings and equipment, qualifies as a physical loss. Eligible growers may borrow up to 100% of the actual losses (not to exceed $500,000).

Market Loss Payments

In the 2008 farm bill, Congress authorized a one-time "market loss payment" program for asparagus growers.[34] The program provides payments to producers currently growing asparagus for revenue losses during crop years 2004-2007 due to imports, totaling $7.5 million for producers of fresh asparagus and $7.5 million for producers of processed or frozen asparagus. Previously, Congress had authorized market loss payments for apple growers: one in each of the FY2001 and FY2002 Agriculture appropriations laws (P.L. 106-387, P.L. 107-76), and one in the 2002 omnibus farm law (P.L. 107-171). These programs provided $269 million for apple grower income assistance in the 1999 and 2000 crop years. Market loss programs are administered by FSA.

Protection for Sellers

The Perishable Agricultural Commodities Act of 1930 (PACA) and the Produce Agency Act of 1937 are the primary laws exclusively serving the produce industry.[35] Under these acts USDA's Agricultural Marketing Service (AMS) administers a program to protect producers, shippers, distributors, and retailers from loss due to unfair or fraudulent practices in the marketing of fresh and frozen fruits and vegetables. PACA was enacted at the request of the fruit and vegetable industry to establish and enforce a code of fair business practices. Under PACA, commission merchants, dealers, and brokers handling perishable agricultural commodities in interstate and foreign commerce must obtain a license and abide by certain fair trading practices.[36] Traders who violate PACA face license suspension or revocation. PACA also provides an administrative dispute resolution process for settling complaints of violations between buyers and sellers.

Congress amended PACA in 1984 to create a statutory trust consisting of a buyer's business-related assets. In the event a buyer fails to make full payment (due to bankruptcy, for example), fruit and vegetable sellers can recover money owed to them before trust assets are made available to general creditors. PACA activities are funded by fees charged for obtaining licenses and for filing complaints. From FY2000 to FY2009, USDA conducted more than 200 enforcement actions to sanction firms and individuals for PACA violations.37 In 2011, AMS resolved a reported 1,563 commercial disputes. Decisions and orders were issued in 427 formal reparation cases involving award amounts totaling approximately $11 million. AMS initiated 17 disciplinary cases against firms for alleged PACA violations and issued 19 disciplinary orders—either suspending or revoking a firm's PACA license, levying civil penalties, or issuing a finding of repeated and flagrant violations against produce firms for violations of the PACA.[38]

Marketing and Promotion

AMS administers several different types of programs intended to help the produce industry expand its markets. AMS's mission is to facilitate the competitive and efficient marketing of agricultural products. These programs include marketing orders and agreements, research and promotion programs, collecting and disseminating USDA's *Market News* reports and information, and providing an array of grading, quality certification, inspection, and product standardization services for fresh and processed produce, among others. AMS administers most of the marketing and promotion programs that benefit specialty crop producers, such as the Specialty Crop Block Grant Program and other programs.

Specialty Crop Block Grant Program

The Specialty Crop Block Grant Program (SCBGP), administered by AMS, was authorized in the Specialty Crops Competitiveness Act of 2004, and further amended by the 2008 farm bill.[39] Under the program, USDA provides block grants to the state departments of agriculture within the 50 states, the District of Columbia, and the U.S. territories to enhance the competitiveness of specialty crops. The program receives mandatory funding through the CCC, available without an annual (or discretionary) appropriation. Program funding will have totaled $224 million over the FY2008-FY2012 period: $10 million (FY2008); $49 million (FY2009); and $55 million annually (FY2010-FY2012).

Under the program, each state receives a base grant plus additional funds based on the state's share of the total value of U.S. specialty crop production.[40] California ($18.7 million), Florida ($4.4 million), and Washington ($3.1 million) have been the three largest recipients under this program, accounting for nearly one-half of all available funds.[41] Several states receive about $1- $2 million each.[42] Most other states and territories receive a total of roughly $200,000 to under $1 million each. How each state spends its allocation depends on its priorities. In FY2011, a total of 739 projects were funded, covering marketing and promotion (33% of projects), pest and plant health (16%), research (15%), education (14%), food safety (9%), production (6%), and other types of projects (7%), as shown in Figure 6.[43] USDA's annual report describes the funded projects across all states.[44]

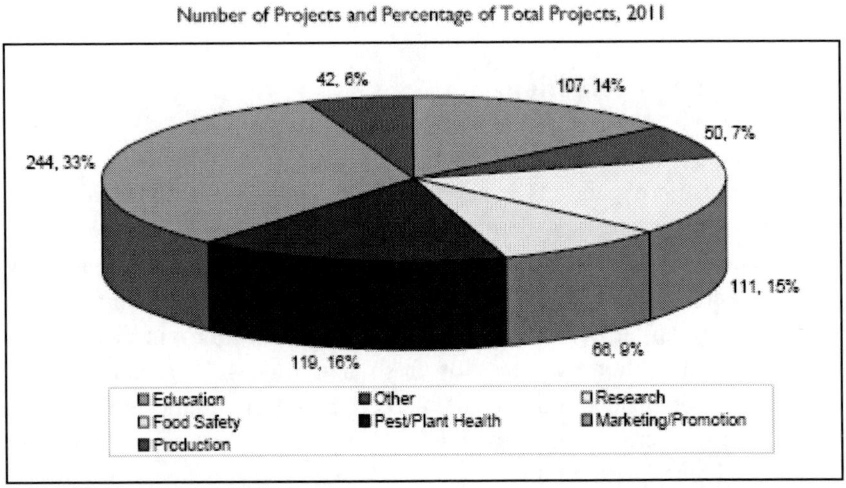

Source: USDA, "Funded Projects," http://www.ams.usda.gov/AMSv1.0/getfile?dDocName= STELPRDC5093992.

Figure 6. Specialty Crop Block Grant Program Projects, by Type.

Value-Added Producer Grants

The Value-Added Producer Grants (VAPG) program was originally authorized by the Agricultural Risk Protection Act of 2000, and amended by subsequent farm bills.[45] The program, administered by USDA's Rural Business-Cooperative Service, provides grants to eligible entities, such as independent agricultural commodity producers, agricultural producer groups, farmer and rancher cooperatives, and majority-controlled producer-based businesses, to develop strategies and business plans to further refine, enhance, or otherwise add value to their products. Grants may be used for planning activities (such as development of feasibility studies, business plans, and marketing strategies) and for working capital to implement a marketing strategy for value-added agricultural products and for farm-based renewable energy. The maximum amount of a planning grant is $100,000 and of a working capital grant is $300,000. Grant funds may be used to pay up to 50% of a project's costs, with the applicant contributing at least 50% in cash or in-kind contributions.[46] Value-added producer grants offer another potential resource for specialty crop growers to engage in market and product development, as well as to finance various value-added activities, such as further processing and packaging of raw agricultural commodities.

Available funding is both mandatory and subject to annual appropriations. Current mandatory funding levels provided $15 million for FY2008, which is available until expended. Discretionary funding is authorized at $40 million annually from FY2008 to FY2012.[47] Since the program began in 2001, the total amount of grant funding provided has ranged from about $15 million to more than $20 million annually. In FY2009, $22.4 million in grants were awarded. A full listing of all FY2011 VAPG recipients is available at USDA's website.[48]

Farmer Direct Marketing Assistance

USDA's farmers' market and various other direct-to-consumer marketing programs provide for market access and assistance to small and medium-size farmers, including fruit and vegetable growers. The intent of the Farmer-to-Consumer Direct Marketing Act of 1976 (P.L. 94-463) was to promote the "development and expansion of direct marketing of agricultural commodities from farmers to consumers" through a range of marketing channels including farmers' markets, farm stands, and roadside stands, community-supported agriculture (CSA), "pick-your-own" farms, Internet marketing, and other types of niche markets. The act originally authorized the Farmers' Market Promotion Program (FMPP), administered by AMS, which was amended in the 2002 and 2008 farm bills.[49]

Under the FMPP, USDA provides grants to establish, improve, and promote farmers' markets and other direct marketing activities. Eligible entities include farmer cooperatives, grower associations, nonprofit/public benefit corporations, local governments, economic development corporations, regional farmers' market authorities, among others. FMPP grants are available to raise market awareness of local foods through promotion, outreach, and advertising; educate farmers and growers in marketing and business planning; and to purchase infrastructure, such as refrigerated trucks, or equipment for a commercial kitchen for value-added products.[50] Grant awards are limited to $100,000, with a minimum award of $5,000. Matching funds are not required. Authorized funding is through the CCC: $3 million (FY2008); $5 million

(FY2009- FY2010); and $10 million annually (FY2011-FY2012). A listing of FY2011 awards is at USDA's website.

FMPP grants are also available to bring local farm products into federal nutrition programs through electronic benefits transfer (EBT) technology at direct-market outlets in order to accept Supplemental Nutrition Assistance Program (SNAP, formerly the food stamp program) benefits. In addition to SNAP, FNS administers two other related programs: the WIC Farmers' Market Nutrition Program (WIC-FMNP)[51] and the Senior Farmers' Market Nutrition Program (SFMNP).[52] These two programs allow for farmers' market purchases by low-income WIC applicants and recipients and also low-income seniors, usually through the use of redeemable coupons. For more information on those programs and redemption at farmers' markets, please see "Assistance to Households and Families."[53]

Market News

The AMS *Market News* program is authorized by various statutes including the Agricultural Marketing Act of 1946 and several omnibus farm bills (1981, 1985, and 2008), among other statutes.[54] Under the program, AMS collects, analyzes, and disseminates local, regional, national, and international market information for many agricultural commodities, including fruits, vegetables, and ornamentals.[55] Federal and state reporters collect data (provided on a voluntary basis) at wholesale markets, farmers' markets, shipping points, and other locations, and also by phone and electronically. AMS disseminates the information on the Internet on a variety of schedules, depending upon the needs of the specific commodity. The information includes supply, prices, contractual agreements, inventories, movement, and more.

The total annual appropriation for *Market News* is approximately $33 million. Of this amount, the 2008 farm bill authorized appropriations of $9 million annually (FY2008-FY2012), "to remain available until expended," to support the collection and dissemination of market news for specialty crops.[56] The 2002 and 2008 farm bills also provided funding to support data collection of certified organic agricultural products (discussed later under "Product and Market Data Collection").

Marketing Orders and Agreements

Marketing orders and agreements are managed by administrative committees made up of local growers and handlers who are operating under them. AMS publishes the proposed and final regulations in the *Federal*

Register. These regulations may include quality standards; quantity controls; grading, certification, and verification; packaging requirements; research and promotion; packaging standards; among other things. Imported products of commodities covered by a marketing order or agreement are also covered. The activities of marketing orders and agreements are financed by industry assessment fees (commonly called "check-off" fees) collected from handlers, usually at the time of sale. To administer the orders and assure that they operate legally and in the public interest, AMS uses funds provided through annual USDA appropriations acts.[57]

The Agricultural Marketing Agreement Act of 1937 authorizes AMS to facilitate and oversee the operation of marketing orders and agreements, usually at the request of industry.[58] Producers and handlers in a specific growing area generally initiate the administrative process leading to the establishment of an order or an agreement. Once a two-thirds majority of the parties in that area approves a marketing *order* by referendum, the order is binding on all growers and handlers in that area. In contrast, a marketing *agreement* is binding only on growers and handlers who are voluntary signatories to the agreement. Currently there more than 20 active marketing orders and agreements. Fruits, vegetables, and nuts covered by federal marketing orders include almonds, apricots, avocados, sweet and tart cherries, citrus in Florida and Texas, cranberries, dates, grapes, hazelnuts, kiwifruit, nectarines, olives, onions (selected types and regions), peaches, pears in Oregon-Washington, pistachios, California and Washington plums/prunes, potatoes in selected areas, raisins, spearmint oil, tomatoes, and walnuts.[59]

In April 2011, AMS issued a proposed rule for a national marketing agreement reflecting USDA and FDA recommendations for food safety practices for leafy greens, as part of a "National Marketing Agreement Regulating Leafy Green Vegetables."[60] Also referred to as the National Leafy Greens Marketing Agreement (NLGMA), the rule would cover the handling of selected leafy greens—spinach, lettuce, and cabbage. It would establish a voluntary program to provide "a governance structure for farmers, handlers, retailers and consumers to work together and develop a practical program so that all types of farming and handling operations can effectively and efficiently comply with food safety requirements."[61] AMS's proposal has been under consideration at USDA for the past few years and reflects an industry-led effort to establish a voluntary program requiring compliance of its signatories (marketing agreement), including importers, in meeting certain commercial food quality and safety requirements. The concept originated with the California Leafy Green Products Handler Marketing Agreement, and covers a range of leafy green products.[62]

USDA's final rule on the NLGMA has not yet been published. It remains unclear how USDA's proposed voluntary efforts for leafy greens would interact with food safety regulations for a wider range of fruits and vegetables that are being developed by the Food and Drug Administration (FDA), as mandated by the FDA Food Safety Modernization Act (FSMA, P.L. 111-353). Some groups also argue that USDA's marketing agreement program could lead to confusion among consumers;[63] others question whether USDA has the expertise and the mandate to regulate food safety.[64] For additional information, see "FDA Food Safety Regulations."

Inspection, Grading, Standardization, and Other Promotion

The Agricultural Marketing Act of 1946 directs USDA to provide such quality grade standards to encourage uniformity and consistency in commercial practices.[65] AMS develops quality grade standards for commodities as needed by the agriculture and food industry, mostly under cooperative agreements with 48 states and Puerto Rico.[66] Under federal-state agreements, AMSlicensed state employees work where needed: in fields during harvest; at land, sea, and air ports of entry; and at packing houses, processing plants, warehouses, and federal and federal-state terminal markets. In FY2011, AMS graded approximately 15.8 billion pounds of processed fruits and vegetables at 381 processing plants, 14 field offices, and 13 inspection points.[67]

Grading is paid for by user fees and is voluntary unless the commodity is regulated for quality under a marketing order or agreement, subject to export requirements, or purchased by USDA or another federal agency for distribution (e.g., through the school lunch program or the military). Shipments of any imported commodity whose domestic production is under a marketing order or agreement must receive AMS grading to assure that the produce is comparable to the U.S. grade, size, quality, and maturity requirements. More than 300 grade standards for fresh and processed fruits, vegetables, nuts and other specialty crops are listed at USDA's website.[68]

Finally, AMS administers several federal commodity research and promotion programs, also known as check-off programs, that have been established at the request of some specialty crop industries.[69] These programs allow farmers, ranchers, and other stakeholders to pool funds and develop a coordinated program of research, promotion, and consumer information to improve, maintain, and develop markets for their products. Specialty crop industries with check-off programs include blueberries, Hass avocados, mangos, mushrooms, potatoes, and watermelons.

Standards of Identity

The Federal Food Drug and Cosmetic Act (FFDCA) directs FDA to establish definitions and standards for food to "promote honesty and fair dealing" for the benefit of consumers.[70] Under the statute, FDA is authorized to establish regulations "for any food ..., a reasonable definition and standard of identity, a reasonable standard of quality, and reasonable standards of fill" of the container for any food. FDA has established roughly 300 identity standards in 20 categories of food, consisting of a range of processed foods and meat, dairy, and seafood products, as well as preserved and processed fruit and vegetable products and juices. Standards of identity cover mostly processed and value-added foods, including canned fruits and vegetables, frozen vegetables, fruit and vegetable juices and beverages, jellies and preserves, tree nut products, and other foods.[71] The statute states that no definition and standard of identity and no standard of quality be established for fresh or dried fruits and vegetables, except for avocadoes, cantaloupes, citrus fruits, and melons.

FDA may initiate the development of a standard in cases where it determines a standard is in the interest of consumers or in response to a petition. The rulemaking process to develop a food standard can be time-consuming, often requires detailed technical expertise, and may generate input by supporters and opponents of the proposed recipes; also, the burden of providing information to support the petition is on the petitioner.[72] The process is similar to those for other FDA rulemaking actions, such as establishing requirements for food additives and ingredients, color additives, and other product claims.[73]

Country-of-Origin Labeling

Country-of-origin labeling (COOL) refers to a labeling law that requires retailers (including grocery stores, supermarkets, and club warehouse stores) to notify their customers with information regarding the source (origin) of certain foods. Originally authorized in the 2002 farm bill, COOL prescribes specific criteria that must be met for a covered commodity—both domestic and imported products—to bear a "United States country of origin" declaration.[74] Covered commodities include many types of specialty crops including fresh and frozen fruits and vegetables, ginseng, pecans and macadamia nuts,[75] among other foods, such as selected meat products, wild and farm-raised fish and shellfish, and peanuts. AMS is responsible for administration and enforcement. The final rule for all covered commodities went into effect on March 16, 2009.[76] USDA estimated that about 86,500 fresh and processed fruit and vegetable, ginseng, and tree nut establishments would be affected by the rule.[77] Reportedly, surveys conducted before the final rule took effect indicated that more than 50% of fresh produce offered for sale in retail grocery stores was labeled with country of origin packaging stickers.[78]

Food Safety

Food safety is a critical issue for the specialty crop industry, as consumers increasingly are recognizing the importance of fruit and vegetable consumption to long-term health and proper weight maintenance. Nonetheless, the nature of production, handling, and preparation makes produce vulnerable to contamination from a wide variety of sources. The fact that produce often is consumed raw contributes to its potential as a source of foodborne illness, attributable in part to the growth in consumer preference for fresh, pre-cut produce, as well as the widespread use of such products in restaurants.

Statistics compiled by the Center for Science in the Public Interest (CSPI) foodborne illness outbreak database indicate that products classified under the "Produce" category in CSPI's database were associated with 639 outbreaks and 31,496 associated illnesses between 1990- 2009.[79] Some of the more recent outbreaks have been attributed to leafy greens, alfalfa and clover sprouts, celery, tomatoes, and green onions. Microbial hazards associated with produce include pathogenic (disease-causing) strains of *Escherichia coli*, *Salmonella*, *Vibrio*, *Shigella*, *Cryptosporidium*, *Giardia*, *Cyclospora*, *Toxoplasma gondii*, and the Norovirus or Norwalk-like virus and Hepatitis A viruses.[80] Also, in 2011, a multi-state outbreak of listeriosis occurred from the contamination of fresh, whole cantaloupe with the pathogen *Listeria monocytogenes*. Such hazards may be introduced during production via agricultural or processing water, soil amendments (manure and municipal biosolids), worker hygiene and sanitary facilities, field and packing facility sanitation, and transportation.

Several federal agencies have oversight responsibility for food safety in the United States.[81] The primary federal agency responsible for produce food safety is the Food and Drug Administration (FDA), within the U.S. Department of Health and Human Services (HHS). Also at HHS, the Centers for Disease Control and Prevention (CDC monitors trends in foodborne illness. Other agencies include the U.S. Environmental Protection Agency (EPA) and the U.S. Customs and Border Protection (CBP). Some USDA agencies also play a role including AMS, as well as the Animal and Plant Health Inspection Service (APHIS), and USDA's research agencies. (This list does not include USDA's Food Safety and Inspection Service (FSIS), which regulates the safety of meat and poultry products, among other animal products.)

FDA Food Safety Regulations

FDA is the primary federal agency responsible for produce food safety, regulating the safety and labeling of all domestic and imported fruit and vegetable products (fresh and processed) and juices and drinks. FDA's authority under the Federal Food, Drug, and Cosmetic Act (FFDCA) was amended by the 111th Congress when it passed comprehensive food safety legislation in the FDA Food Safety Modernization Act, or FSMA.[82]

Under FSMA, FDA is developing mandatory food safety regulations and traceability requirements affecting farmers, packers, and processors of both domestically produced and imported products. At the farm production level, requirements under FSMA § 105 will mostly affect produce growers. Most other types of food producers—such as meat, poultry and dairy farms; fisheries; and producers of raw, bulk grains—will likely not be subject to FSMA farm-level requirements. FSMA also exempted from regulation most small grower and processing operations that sell products locally.[83] Requirements under FSMA § 105 must be established within two years of enactment, and include science-based, minimum standards for the safe production and harvesting of fruits and vegetables. These standards could address certain farm practices at produce operations, including the use of soil amendments, hygiene, packaging, temperature controls, animals in the growing area and water.[84] FDA's rules for these and other requirements are still pending, and reportedly have been at OMB for review.

FSMA requirements that could also affect specialty crop producers include food safety requirements for food facilities (FSMA § 103), which could include new mandatory requirements for produce manufacturers. This rule is also under development by FDA.[85]

FDA also has responsibility for ensuring the safety of imported food, including imported produce. Traditionally, FDA has inspected only 1% to 2% of all annual food imports. FSMA requirements pertaining to all FDA-regulated imports, including produce (FSMA § 301) are also being developed. In addition, following the events of September 2001, Congress passed a bioterrorism preparedness law that addresses import safety (among many other issues). The Bioterrorism Act[86] contains provisions requiring foreign and domestic food establishments to register with FDA and keep thorough records of their purchases and sales, and requiring foreign firms exporting food to the United States to give FDA prior notification of the exact time, location, and contents of incoming shipments.

Other federal agencies play a role in ensuring the safety of imported foods, including CBP, which inspects imported foods, plants, and animals, as well as APHIS, which conducts border inspections, and aims to prevent the introduction or dissemination of plant pests and diseases.

USDA Product Quality and Data Collection Programs

National Marketing Agreement Regulating Leafy Greens

After FSMA was enacted, AMS published a proposed rule in April 2011 to develop and implement USDA-administered requirements, as part of a National Marketing Agreement Regulating Leafy Green Vegetables.[87] This proposed rule covers the handling of fresh leafy green vegetables—spinach, lettuce, cabbage—only. It remains unclear how USDA's proposed voluntary efforts for leafy greens would interact with FDA's rulemaking process to develop mandatory safety standards for a wider range of fruits and vegetables subject to FSMA. For other information, see "Marketing Orders and Agreements."

Qualified Through Verification Program

Since 1996, AMS has offered a voluntary, user-fee, audit-based inspection service for producers of fresh-cut fruits and vegetables to assist produce packers in adopting science-based, preventive measures against food contamination in their plants. The Qualified Through Verification (QTV) program is similar in approach to the preventive Hazard Analysis and Critical Control Point (HACCP) system used by USDA's meat and poultry regulatory agency, the Food Safety and Inspection Service (FSIS).[88] Although the QTV program relates to the safety of fruits and vegetables from a public health standpoint, it is not a regulatory program.

Microbiological Data Program

AMS has administered the Microbiological Data Program (MDP) since 2001.[89] MDP is a national food-borne pathogen monitoring program, implemented with the cooperation of state agriculture departments and other federal agencies, that manages the collection, analysis, data entry, and reporting of foodborne pathogens on selected agricultural commodities.[90] Under the program, fresh produce is tested for the prevalence of harmful bacteria, such as *Salmonella* and pathogenic *E. coli*. Among the types of tested produce is cantaloupe, cilantro, green onions, hot peppers, lettuce, spinach, sprouts and tomatoes. Approximately 17,000 samples have been collected from more than 600 food distribution sites under the program.

According to USDA, FDA is notified whenever a product tests positive for the presence of harmful bacteria and the source of contamination can be removed from the food distribution system. CDC also is notified to aid in the surveillance of food-related outbreaks. Opinions differ on the value of the program, and some groups have sought to eliminate or to redesign the program: Some have expressed concern that such a food safety program does not belong at USDA and should be housed at FDA; other groups maintain that the program provides useful information to consumers and are further concerned that FDA might not have the resources and expertise to maintain the program.[91] In recent years, USDA has spent between \$4 million to \$5 million annually to operate the program.[92]

Pesticide Residues

EPA is responsible for regulating pesticide use on food and determining whether and under what conditions the proposed pesticide use would present an unreasonable risk to human health or the environment. When Congress enacted the Food Quality Protection Act of 1996 (FQPA), it established a new standard of safety for pesticide residues on food. Maximum pesticide residue levels (known as "tolerances") must be set by EPA to ensure with "a reasonable certainty" that "no harm" will come to children as a result of pesticide exposure.[93] EPA regulates the labeling, sale, and use of pesticides on domestically produced and imported food toward that safety goal. FDA is responsible for ensuring that tolerance levels for food are not exceeded. Based on the data submitted by pesticide manufacturers when they apply to register a pesticide active ingredient, pesticide product, or a new use of a registered pesticide under Section 3 of the Federal Insecticide, Fungicide, and Rodenticide Act (FIFRA), EPA determines whether and under what conditions the proposed pesticide use would present an unreasonable risk to human health or the environment. If the pesticide is proposed for use on a food crop, EPA also determines whether a "safe" level of pesticide residue, called a "tolerance," can be established under FFDCA.

In cooperation with EPA, FDA determines which pesticides, insecticides, fungicides, and herbicides may be used on fruit and vegetable crops, and what chemical residue levels will pose the least risk to human health at normal consumption rates. FDA regulations impose the same standards on countries that export produce to the United States, and the agency is responsible for inspecting imports for safety.

At USDA, AMS administers a cooperative federal-state residue testing program through which it collects data on residual pesticides, herbicides, insecticides, fungicides, and growth regulators in over 50 different commodities, including fresh/frozen/canned fruits and vegetables, and fruit juices, among other things.[94] The pesticides and commodities to be tested each year are chosen based on EPA data needs, and on information about the types and amounts foods consumed, in particular, by infants and children. The Pesticide Data Program (PDP) is a national pesticide residue database program that collects data from fresh, frozen, and canned fruits and vegetables, fruit juices, and nuts, among other foods (domestic and imported) at more than 600 sites in 11 participating states.[95] In FY2005, more than 11,000 fresh and processed produce samples were tested under the program.[96]

Export and Trade Promotion

USDA trade promotion programs, such as the Market Access Program (MAP) and other market development programs, support many export-oriented markets within the specialty crops and certified organic agriculture. Other trade remedy programs are also available. These programs are mostly administered by USDA's Foreign Agricultural Service (FAS).

Market Development Programs

The Market Access Program (MAP) was established to facilitate U.S. agricultural exports, as part of the Agricultural Trade Act of 1978, as amended by subsequent farm bills.[97] The program uses CCC funds to help U.S. producers, exporters, private companies, and other trade organizations finance promotional activities for U.S. agricultural products. MAP (formerly the Market Promotion Program) encourages the development, maintenance, and expansion of commercial export markets for agricultural commodities through cost-share assistance to eligible trade organizations that implement a foreign market development program. Activities financed include consumer promotions, market research, technical assistance, and trade servicing. MAP money can be used to support both brand-name promotions and generic promotions.[98] The program is administered by Foreign Agricultural Service (FAS).

MAP is widely used by some specialty crop growers to encourage domestic exports. The 2008 farm bill also specifically added language to address coverage for certified organic foods.[99] Mandatory funding, as authorized by the 2008 farm bill, is $200 million annually through FY2012 for all overseas agricultural promotion and marketing activities. Of this amount, in FY2010, about 30% of all MAP funding—nearly $60 million—went to specialty crop producer groups in FY2010.[100] Nearly half supported California specialty crop groups in the almond, asparagus, cherry, citrus, kiwifruit, peach, pear, pistachio, prune, strawberry, table grape, tomato, tree fruit, and walnut sectors. Approximately $8 million supported organizations of the U.S. wine industry. A total of about $16 million went to groups supporting specialty crops in Florida, Texas, Hawaii, Washington, and some northwestern states in FY2010. Another roughly $8 million supported national groups in the apple, cherry, cranberry, potato, and watermelon sectors. In FY2010, the Organic Trade Association received $0.4 million under MAP.

FAS administers other trade development programs that support certain U.S. specialty crops.

- The Quality Samples Program (QSP) helps create export sales of commodities by providing samples to foreign importers, thus paving the way for new partnerships between importers and U.S. exporters.101 Total FY2010 (mandatory) funding for the program was $1.9 million, of which about 40% of funds were directed toward specialty crop groups (cranberry, ginseng, potato, and walnut samples to potential importers). This amount does not include support through other national, state, or regional export promotion groups that might also provide support for specialty crops, among other agricultural commodities.
- The CCC export credit guarantee program promotes purchases of U.S. agricultural exports by providing competitive credit terms to foreign buyers through credit guarantees.102 CCC funds guarantee the payments due from approved foreign banks to U.S. exporters or financial institutions. The CCC determines which countries and banks are eligible and at what level of debt, and also selects which commodities and products will be eligible (depending upon market potential). Among the eligible U.S. agricultural commodities are a

wide variety of fresh, dried, and processed fruits; canned, dried, fresh, and frozen vegetables; juices; tree nuts; wine; and nursery products.103

Additionally, FAS administers other programs, such as the Foreign Market Development (FMD) program and the Food for Progress (FFP) program. FMD (also known as the cooperator program) provides for cost-sharing of overseas marketing and promotion activities with nonprofit U.S. commodity and trade organizations.[104] This program received mandatory CCC funding for overseas promotion of U.S. bulk commodity crops, but not for fruits, vegetables, and tree nuts.[105] FFP provides for the donation of U.S. agricultural commodities to certain developing countries, whereby donated commodities are monetized (sold on the local market) and the proceeds are used to support agricultural development activities.[106] Previously FFP had procured small quantities of U.S. dehydrated potatoes and dehydrated vegetables. In 2009, only dehydrated potatoes comprised an overall very small share of USDA's overseas food aid under the program.[107]

Technical Assistance for Specialty Crops

The Technical Assistance for Specialty Crops (TASC) program, administered by FAS, was originally authorized in the 2002 farm bill and reauthorized in the 2008 farm bill.[108] The program provides funds to eligible entities for projects that address sanitary (animal) and phytosanitary (plant) barriers, commonly referred to as SPS barriers to U.S. specialty crop exports. SPS requirements and their potential to be trade barriers, among other types of technical barriers to trade (TBT), has become a more prominent issue as tariffs have been reduced under multilateral trade agreements and various free trade agreements (FTAs) entered into by the United States, such as the North American Free Trade Agreement and other bilateral FTAs.[109] A summary of the current U.S. concerns regarding SPS and TBT issues across all agricultural commodities and U.S. trading partners is provided in annual reports compiled by the Office of the United States Trade Representative (USTR).[110]

TASC projects should demonstrably benefit the represented industry rather than a specific company or brand, and must address barriers to exports of commercially available U.S. specialty crops for which barrier removal would predominantly benefit U.S. exports. Examples of expenses that CCC may agree to reimburse under the TASC program may include initial pre-clearance programs, export protocol and work plan support, seminars and workshops, study tours, field surveys, development of pest lists, pest and disease research, database development, reasonable logistical and administrative support, and travel and per diem expenses. Eligible recipients include U.S. federal or state government agencies, U.S. nonprofit trade associations, U.S. universities, U.S. agricultural cooperatives, U.S. private companies or any other U.S. organizations. Funding is through the CCC, and was authorized at $4 million (FY2008), $7 million (FY2009), $8 million (FY2010), and $9 million annually (FY2011 and FY2012). The program is administered by FAS.

Trade Adjustment Assistance for Farmers

The Trade Adjustment Assistance (TAA) for Farmers program provides technical assistance and cash benefits to eligible producers of agricultural commodities and fishermen who experience adverse economic impacts caused by increased imports.[111] TAA for Farmers provides technical assistance and cash benefits to eligible farmers and fishermen who have been adversely affected by competition from imports of a commodity that they produce,[112] if increased imports have contributed importantly to a price decline of at least 20%. Support is available in the form of enhanced technical assistance and seed money to enable a producer to formulate and implement a business adjustment plan. The program is administered by FAS.

Under the program, fish and seafood producers have accounted for most of the cash benefits paid out. Among fruit and vegetable growers, producers of Concord grapes, lychees, olives, wild blueberries, fresh potatoes, Florida avocadoes, and asparagus were among others that USDA certified to be eligible for assistance. Funding is discretionary, and currently may not exceed $90 million annually (FY2012-FY2013), and $22.5 million (first quarter FY2014).

Trade Remedies

In the event of suspected unfair competition from foreign imports, U.S. law makes available certain remedies that the specialty crop industry can pursue, not within USDA, but from the Department of Commerce and the U.S. International Trade Commission. Title VII of the Tariff Act of 1930 provides for the levying of antidumping (AD) duties on imports sold at less than fair value that have caused or threaten to cause material injury to a domestic industry producing a like product. Where subsidized imports have this injurious effect, Title VII authorizes countervailing duties (CVD) to be imposed.[113]

U.S. specialty crop producers on occasion have petitioned the Department of Commerce and the USITC to investigate suspected occurrences of dumping. Previous USITC investigations have highlighted the increased competitive market and trade pressures on U.S. fruit producers from lower-cost foreign fruit and vegetable producers (such as those in China, Thailand, Chile, Argentina, and South Africa) as well as from countries with subsidized fruit and vegetable production (such as in the EU, including Spain).[114] Import injury investigations initiated by the United States further highlight concerns that some countries might be supplying imports at prices below fair market value. Since the 1990s, dumping petitions filed by the U.S. fruit and vegetable sectors have included charges against imports of fresh tomatoes (Canada, Mexico), frozen raspberries (Chile), apple juice concentrate (China), frozen orange juice (Brazil), lemon juice (Argentina, Mexico), fresh garlic (China), preserved mushrooms (China, Chile, India, Indonesia), canned pineapple (Thailand), table grapes (Chile, Mexico), and tart cherry juice (Germany, former Yugoslavia).[115] Many of these petitions were decided in favor of U.S. domestic producers and resulted in higher tariffs being assessed on U.S. imported products from some of these countries.

Pest and Disease Exclusion

USDA's Animal and Plant Health Inspection Service (APHIS) is responsible for protecting U.S. agriculture from domestic and foreign pests and diseases, responding to domestic animal and plant health problems, facilitating agricultural trade, regulating genetically engineered organisms, and other responsibilities related to animal welfare and wildlife damage management.[116] For the fruit and vegetable industries, APHIS addresses pest and disease exclusion (i.e., prevention, detection, and eradication) as well as emergency response, management, trade issue resolution, and capacity building.

The Plant Protection Act (PPA) is the primary law governing APHIS's responsibilities regarding plant health consolidated several plant quarantine authorities, some dating back to the 1880s.[117] (The primary law governing animal health is the Animal Health Protection Act, AHPA). PPA authorizes APHIS to cooperate with states, localities and others to prevent the spread of and eradicate invasive pests and diseases. The statute also authorizes APHIS to prohibit or restrict the importation, exportation, and the interstate movement of plants, plant products, certain biological control organisms, noxious weeds, and plant pests, and also authorizes APHIS to inspect foreign plant imports, to quarantine any state or premises infested with a new pest or noxious weed, and to cooperate with states in certain control and eradication actions. These authorities have been traditional hallmarks of U.S. plant pest regulations, and are administered by APHIS in collaboration with state departments of agriculture and their plant protection boards.

PPA gives USDA authority to use a wide range of measures to exclude alien pests or prevent the spread of new, but not widespread pests. These measures include inspections, surveillance, quarantines, treatments, or destruction. USDA can develop lists of organisms that can or cannot enter the United States and goods that can be imported from specific countries, and has the authority to certify that U.S. agricultural exports meet the phytosanitary standards of other countries. USDA can require private parties to take remedial actions without cost to the government but must select the least costly, effective measure. USDA has less regulatory authority to address established and widespread pests, but can enter into agreements with foreign governments, state governments, or other organizations to implement the act.

For the most part, APHIS, which has a nationwide network of regional and state offices, serves in a consultative mode to assist state departments of agriculture in planning and operating control and eradication programs using state and private funds. However, when a particularly harmful disease or pest emerges suddenly, state resources for immediate response can be quickly overwhelmed.

In such emergency situations, USDA has broad authority to transfer funds from the Commodity Credit Corporation (CCC) to APHIS for emergency control programs. The authority to transfer money for plant and animal health emergencies is found both in annual appropriations acts and in authorizing statutes, including the PPA.[118] Such authorities date back to 1948. Discretion rests with the USDA Secretary, who is subject to limited review when making transfers. In recent appropriations, appropriators have expressed the expectation that USDA will continue to use its authority to transfer funds from other appropriations or funds available to USDA for activities related to the arrest and eradication of animal and plant pests and diseases. USDA has exercised this authority in recent years, and it has become an issue within government concerning the method for funding plant and animal health programs. USDA reports that in FY2011 it redirected $65.9 million in emergency funding for activities covering some plant-related concerns caused by the Asian longhorned beetle, European grapevine moth, and the light brown apple moth, among other pests.[119]

Pest Detection and Surveillance

The 2008 farm bill amended PPA to provide for early plant pest detection and surveillance, threat identification and mitigation of plant pests and diseases, and technical assistance in the development and implementation of audit-based certification systems and nursery plant pest risk management systems ("section 10201").[120] At the same time, Congress also established a related program, the National Clean Plant Network (NCPN), to provide reliable sources of pathogen-free planting stock of high-value specialty crops.[121] According to USDA, certain aspects of these programs connect both programs.[122] Funding is mandatory through CCC. Section 10201 received authorization for $12 million (FY2009); $45 million (FY2010); and $50 million (FY2011- FY2012 and each fiscal year thereafter). NCPN was provided with $5 million annually (FY2009- FY2012).

APHIS is implementing Section 10201 across six goal areas: (1) enhancing plant pest/disease survey and analysis; (2) targeting domestic inspection activities at vulnerable points;

(3) enhancing pest identification tools and technology; (4) developing programs to safeguard nursery production; (5) enhancing outreach and education; and (6) enhancing mitigation capabilities. In FY2011, APHIS funded 312 projects across all goal areas. About three-fourths of the projects directly provided funds to 48 state departments of agriculture and two territories. The remaining one-fourth of all projects provided funds to universities, federal agencies, tribal organizations, and nonprofit entities. Funds were used by APHIS for certain programs, including development of an improved data management system. More detailed information is available in USDA's FY2013 budget justification.[123]

APHIS, ARS, and NIFA are working to develop the NCPN under a memorandum of understanding. As of 2012, APHIS has entered into 19 cooperative agreements with clean plant centers.[124] Centers use NCPN funds to (1) diagnose for harmful pathogens that cause disease in covered specialty crops, (2) apply therapeutic measures to eliminate these pests, (3) establish plantings of clean plant 'starter' material and make this material available to nurseries and growers, and (4) engage with nurseries and growers in education/outreach programs to communicate the economic value to industry of using clean nursery stock. These activities are expected to provide additional sources of healthy planting stock for fruit trees, grapes, citrus, berries, and hops.

The 2008 farm bill also authorized "such sums as necessary" to establish a Pest and Disease Revolving Loan Fund to provide loans to local governments to finance purchases of equipment to monitor, remove, dispose of and replace pest- and disease-infested trees in quarantine areas. The Forest Service is drafting rules and identifying appropriate mechanisms to implement the fund.[125]

Specialty Crop and Plant Pest Management

APHIS is asking to spend an estimated $151.1 million to address specialty crop pests in FY2013, roughly 20% of its total appropriation for that year.[126] Much of this is allocated to APHIS's Emerging Plant Pest (EPP) program. Under EPP, APHIS cooperates with states to develop, implement, and fund action plans for surveying, reporting, and controlling emerging pest threats. EPP provides APHIS with the infrastructure to carry out urgent plant pest and disease programs, some of which currently are or have been partially funded through emergency CCC transfers.

For specialty crops, EPP is working to address concerns within the following areas: Citrus Health Response Program, Asian longhorned beetle, emerald ash borer, glassy-winged sharpshooter, pale cyst nematode, and light brown apple moth.

Additionally, APHIS gets appropriated funds for its Pest Detection program.[127] The program helps ensure that any new introductions of harmful plant pests and diseases are detected as soon as possible, before they cause significant damage. Under the program, APHIS works with its state cooperators and also the scientific community, universities, the public, non-profit entities, and industry to carry out surveys for high-risk pests, diseases, and weeds in the field. Funding is provided through the Cooperative Agricultural Pest Survey (CAPS) program. Information collected through CAPS is compiled into detailed maps and other formats, and filed in the electronic National Agricultural Pest Information System (NAPIS) database. The program helps identify pest-free regions and allow for continued export of commodities from particular areas of the country. When significant quarantine pests are found, APHIS and cooperators rapidly decide an appropriate course of action.

The CAPS/NAPIS system allows for early detection of significant pests, which in turn helps organize eradication efforts before pests cause major economic damage. These efforts also support inspections of commodities, conveyances, and passenger baggage conducted by CBP at sea ports, airports, and land border crossings. APHIS is asking to spend an estimated $25.6 million to address pest detection across all crops and program areas in FY2013.[128]

Import Inspection and Quarantine

APHIS and the Department of Homeland Security's (DHS's) Bureau of Customs and Border Protection (CBP) administer the Agricultural Quarantine Inspection (AQI), which protects the United States from the risks associated with the introduction of invasive agricultural pests and diseases.[129] Under the program, APHIS and CBP administer foreign plant quarantines, whereby the importation of certain plants and plant products into the United States may be prohibited or restricted.[130] APHIS inspects passengers, cargo, and conveyances traveling from Hawaii, Puerto Rico, and other islands to the mainland. Among APHIS's pest and disease exclusion activities are to (1) develop protocols for plant materials in trade; (2) maintain quarantine facilities and treat regulated imported products; (3) conduct pre-clearance programs for products being imported into the United States and certification programs for U.S. agricultural exports; and (4) support scientific projects to detect and identify high-risk plant pathogens, and develop protocols for quarantine testing.[131] The program is funded through a combination of appropriations and user fees. APHIS expects it will spend about $25 million for the APHIS AQI program in FY2013 (covering all U.S. agriculture).[132] Within APHIS, these activities are carried out under the agency's Plant Protection and Quarantine (PPQ) program.

The Center for Plant Health Science and Technology (CPHST), located on USDA's research campus in Beltsville, MD, is a key component of the APHIS's National Plant Pathogen Laboratory Accreditation Program (NPPLAP). The lab is responsible for proficiency test panel development, delivery, and first-level evaluation of proficiency tests conducted by scientists who perform diagnostics on behalf of APHIS using CPHST-validated methods.[133] In addition, APHIS maintains a searchable database, known as the Fruits And Vegetables Import Requirement (FAVIR) database, which provides an online reference to regulations and information pertaining to the importation of fruits and vegetables into the United States.[134] APHIS also maintains information on specific agricultural pests and diseases,[135] among other types of program activities.

Export Facilitation

Along with overseeing U.S. import requirements, APHIS also provides assistance to U.S. specialty crop growers who have the capacity to export crops abroad. APHIS helps to maintain and expand existing markets and create new markets, assisting U.S. exporters to meet the entry requirements of other countries and also resolve trade issues to facilitate U.S. exports, and also building international support for trade agreements.[136] APHIS attachés, located at U.S. embassies abroad, work with host country officials to establish and oversee foreign-based inspection programs to ensure that products designated for export to the United States are pest-free, and that inspection officials at U.S. ports of entry receive early warning of pest and disease problems that may be emerging in exporting countries. APHIS helps manage and resolve sanitary (animal) and phytosanitary (plant), or SPS, barriers to U.S. exports to other countries (see previous discussion "Technical Assistance for Specialty Crops").

As part of its responsibilities, APHIS, along with other U.S. agencies, represents the United States in the World Trade Organization (WTO) and other international bodies that set SPS standards for trade, and is the USDA negotiator in WTO phytosanitary disputes that concern U.S. agricultural trade. APHIS also helps negotiate and resolve SPS and other types of technical barriers to trade that could potentially affect U.S. trade relationships. USDA reports that through the resolution of SPS issues, APHIS "successfully negotiated trade issues that contributed to the opening of new markets, and retention and expansion of existing markets valued at a total of approximately $2.75 billion (estimated) in 2011."[137] APHIS also is the agency in charge of certifying that U.S. specialty crop exports meet other countries' phytosanitary regulations before they are shipped.[138]

Research and Cooperative Extension

USDA's research and extension service play an important role in specialty crop and organic production through programs directed specifically at specialty crops, as well as general research and extension services available to all U.S. agricultural producers.

The United States has a nationwide network of public agricultural laboratories and academic institutions supported in full or in part by annual USDA appropriations. There are four USDA Research, Education, and Economics (REE) agencies: Agricultural Research Service (ARS); National Institute of Food and Agriculture (NIFA), Economic Research Service (ERS), and National Agricultural Statistics Service (NASS).

- ARS is USDA's chief scientific in-house research agency, and provides scientific and technical support for USDA's regulatory agencies, including APHIS. ARS conducts basic and applied research on the full range of subjects important to specialty crops, from production through processing and food safety. ARS also is the designated lead agency for federal nutrition research.

- NIFA is the USDA agency that distributes federal funds to support research and extension programs at the land grant colleges of agriculture in every state.[139] NIFA supports research, education, and extension programs in the Land-Grant University System and other partner organizations; it does not perform actual research, education, and extension but instead helps fund programs at the state and local level. NIFA allocates some funds to each state according to formulas spelled out in authorizing laws, and distributes the rest through various competitive grant programs.
- ERS is USDA's economic research agency, covering agriculture, food, natural resources, and rural development issues. The agency publishes market analysis and outlook reports for most commodities including specialty crops.
- NASS is USDA's principal data collection agency. In addition to periodic data publications and special reports, NASS also conducts the U.S. *Census of Agriculture* every five years, and conducts the *Census of Horticultural Specialties* once every 10 years. The latter provides the only comprehensive and detailed data compilation of U.S. fruit, vegetable, tree nut, floriculture, nursery, and other specialty crop operations.[140]

This report covers selected USDA research programs that directly support U.S. specialty crop growers, and does not address other research and extension services that generally support all agricultural producers. For example, NIFA is the federal partner in the Cooperative Extension System that provides federal funding to support state, local, and regional offices at land-grant colleges and universities in each U.S. state and territory. These offices are staffed by experts who provide practical and research-based information to agricultural producers, small business owners, and the public. NIFA's website provides contact information and a map of the land-grant colleges and universities across all states and territories.[141] Other NIFA grant programs may also provide indirect support. For example, some specialty crop and organic producers generally benefit from other programs intended to assist farmers in developing and implementing sustainable and innovative farming strategies, such as Sustainable Agriculture Research and Education (SARE) grants through USDA NIFA, and also information services through the National

Sustainable Agriculture Information Service (known as the ATTRA project).[142]

Specialty Crop Research Initiative

The Specialty Crop Research Initiative (SCRI) was authorized in the 2008 farm bill, to provide grants to solve critical industry issues through research and extension activities.[143] SCRI gives priority to projects that are multistate, multi-institutional, or trans-disciplinary; and includes explicit mechanisms to communicate results to producers and the public. Projects must address at least one of five focus areas: research in plant breeding, genetics, and genomics to improve crop characteristics; efforts to identify and address threats from pests and diseases, including threats to specialty crop pollinators; efforts to improve production efficiency, productivity, and profitability over the long term; new innovations and technology, including improved mechanization and technologies that delay or inhibit ripening; and methods to prevent, detect, monitor, control, and respond to potential food safety hazards in the production and processing of specialty crops. The 2008 farm bill provided a total of $230 million in mandatory CCC funds ($30 million in FY2008; $50 million annually for FY2009-FY2012), plus annual appropriations of $100 million (FY2008- FY2012). A listing of funded projects is available at USDA's website.[144]

Methyl Bromide

Through its "National Program 308" (NP 308), ARS is the primary federal research agency conducting research on alternatives to the use of methyl bromide (MeBr)—a pesticidal gas widely used in specialty crop production as a soil fumigant and structural fumigant to control pests use.[145] NP 308 was initiated after methyl bromide was listed as a stratospheric ozone depletor, which was followed by worldwide controls on production, emissions, and trade under the 1987 Montreal Protocol on Substances that Deplete the Ozone Layer, and domestically under Title VI of the U.S. Clean Air Act.[146] The program followed the 1995 formation of the Methyl Bromide Alternatives Working Group to track and facilitate adoption of alternatives to methyl bromide, and the allocation of ARS funds starting in FY1999 toward research to develop alternatives to methyl bromide, principally at the University of California and the California Strawberry Commission, and the University of Florida.

Under the Montreal Protocol, MeBr has been officially phased out as of January 1, 2005. Allowable exemptions to the phase-out include an exemption for Quarantine and Preshipment (QPS) to eliminate quarantine pests, as well as exemptions for *critical use*, or so-called Critical Use Exemptions (CUEs), of which agricultural production is one.[147] CUEs were designed for agricultural users with no technically or economically feasible alternatives to using MeBr. Strawberries, tomatoes, peppers, eggplant, cucurbits,[148] and ornamental nursery crops are particularly dependent on pre-planting soil fumigation with MeBr.[149] Other commodities rely on MeBr to control pests in storage, among other types of post-harvest uses in food processing. Each year some specialty crop growers seek exemptions for critical use in part because research suggests the continued difficulty of finding comparably effective alternatives. Currently, many signatories of the protocol have further agreed to fully phase out MeBr by January 1, 2015, with no provision to exempt critical uses.[150]

Nutrition and Food Assistance

USDA's Food Nutrition Service (FNS) is administers a range of domestic nutrition and food assistance programs. The major laws governing these programs are the Richard B. Russell National School Lunch Act; the Child Nutrition Act; Section 32; the Food Stamp Act, the Emergency Food Assistance Act; and Section 5 of the Agriculture and Consumer Protection Act of 1973.[151]

Congressional jurisdiction over these laws in the Senate is exercised by the Senate Agriculture, Nutrition, and Forestry Committee. However, in the House of Representatives, the jurisdiction is split between the House Education and the Workforce Committee, and the House Agriculture Committee.

These programs do not purchase or benefit the fruits and vegetable industry exclusively, but many aspects of the programs do benefit the industry or have potential to do so.

Commodity Procurement for Domestic Food Assistance Programs

Numerous food distribution programs administered by FNS provide children and low-income individuals access to food and nutrition by providing both funding and USDA-purchased commodity foods. USDA purchase and donation of commodity foods provides food to needy populations, while at the same time supports U.S. agricultural producers, including fruit and vegetable growers. Many of these programs grew out of the programs supporting U.S. agriculture during the Depression.

As part of the USDA Foods program, FNS, AMS, and FSA work together to directly purchase commodities—including fruit, vegetable, and tree nut products—for distribution or donation to various organizations, including schools that provide federally-supported meals.[152]

FNS is responsible for general oversight, regulation, and administration of domestic USDA foods program, is the primary liaison between USDA and the administering state agency, and also tracks entitlement and takes food orders from states.

AMS and FAS are responsible for purchasing and delivering USDA-purchased foods.[153]

There are two types of USDA-purchased food commodities.

- *Mandated, or "entitlement" commodity purchases.* Entitlement commodities refers to food purchases and donations that, by law, *must* be purchased and to which schools, organizations, or states (depending on the program) are entitled. USDA generally purchases entitlement commodities based on preferences expressed by recipient organizations (e.g., schools, state food assistance or program operators).
- *Contingency, or "bonus" commodity purchases.* Periodically, USDA taps its contingency reserve for so-called emergency surplus removals (or diversions), which are then distributed as "bonuses" to domestic food assistance programs. Bonus buys normally are based on market conditions, may be influenced by surpluses or other economic problems with the farming community, and are often intended to stabilize market conditions. In the case of specialty crops, bonus buys tend to include types of fruits, vegetables, and tree nuts not routinely seen on lists of entitlement purchases (e.g., asparagus, apricots, blackberries, almonds).[154]

USDA directly purchases and then donates a variety of non-price-supported commodities, including specialty crops, for consumption through domestic nutrition and food assistance programs.

These purchases and donations help feed groups of nutritionally vulnerable recipients and organizations that serve these groups (such as low-income school children, and participants at family child care homes, child care centers, Head Start programs, and adult care centers, among others) while also helping to balance supply and demand for various commodities.

Federal programs that receive USDA Foods include:[155]

- individuals and household programs, such as the Commodity Supplemental Foods Program (CSFP), The Emergency Food Assistance Program (TEFAP), the Food Distribution Program on Indian Reservations (FDPIR), and disaster feeding programs; and
- schools and institution programs, such as the National School Lunch Program (NSLP), Summer Food Service Program (SFSP), Child and Adult Care Food Program (CACFP), and Nutrition Services Incentive Program (NSIP, formerly Nutrition Program for the Elderly).[156]

Most funding for USDA commodity purchases is classified as "mandatory"—that is, the level is dictated by underlying law. (For example, child nutrition programs are due a specific number of cents per meal in commodity foods.) A lower level of spending is "discretionary"—the amount is set by appropriations decisions or dependent on market conditions. Primary funding sources for USDA commodity procurement include Section 6 of the Richard B. Russell National School Lunch Act; Section 32 of the Act of August 24, 1935 ("Section 32");[157] and Section 416 of the Agricultural Act of 1949.

Depending on the year, roughly 180 food items may be available, including fresh, frozen, packaged, canned, dried, and bulk foods. USDA purchases of fruit and vegetable products represent a significant share of all USDA food purchases. USDA reports that fruit and vegetable purchases by AMS were valued at $613.7 million (FY2010) and $594.3 million (FY2009).[158]

Of total USDA food donation assistance, USDA purchases of fruits and vegetables accounted for more than one-half of the value of all USDA entitlement purchases. The 2002 and 2008 farm bills established minimum levels of specialty crop purchases under Section 32. Minimum purchases for fruits, vegetables, and other specialty crops under Section 32 totaled $406 million in FY2012.[159] In addition, special rules relate to *fresh* fruits and vegetables to child nutrition programs. Under provisions in the 2002 and 2008 farm bills, at least $50 million worth of fresh fruits and vegetables must be provided annually through an arrangement with a Department of Defense (DoD) procurement agency (the Defense Supply Center in Philadelphia). (The initiative is named the Department of Defense Fresh Fruit and Vegetable Program or "DoD Fresh."[160]). The amount is based on the dollar value of commodities that child nutrition programs are entitled to.

Year-to-date AMS purchases in FY2012 (as of August 10, 2012) indicate that fruits and vegetables account for nearly 40% of all purchases (*Table 3*) shows the types of fruits and vegetables purchased over the most recent period under the program. Limited data on the value of USDA bonus buys indicate that fruit and vegetable purchases totaled $1.3 billion over the period from FY2000-FY2009, not including the amount of cash reimbursement to states[161] (*Table 4*). In the case of commodity food assistance programs, USDA reports that of total USDA purchases for child nutrition programs by food type—including both entitlement purchases and bonus buys— fruits and vegetables accounted for 27% in FY2009.[162]

AMS also provides purchasing services to FNS to supply food to recipients in nutrition assistance programs and is reimbursed for the administrative costs associated with these purchases.[163]

Purchases Using Child Nutrition Programs' Cash Assistance

In addition to USDA's purchase of fruits and vegetables as part of the Department's commodity procurement for and donation to domestic food assistance programs, USDA also provides cash assistance based on per-meal reimbursements for the child nutrition programs. "Child nutrition programs" is a category used to describe FNS programs that help to provide food for children in school or institutional settings. The National School Lunch and School Breakfast programs provide a per-meal subsidy for each meal that is served for free, for a reduced-price, or for a full-price (called a "paid" meal). The Child and Adult Care Food Program (CACFP) and Summer Food Service Program (SFSP) will, under certain circumstances, provide free meals or snacks to all the children at a site, because it is the site (not the child) that is subject to eligibility criteria.[164]

These federal funds provide an additional—and proportionally larger—means for institutional purchasing of all foods, including fruits and vegetables. In the case of NSLP, in FY2011, the value of federal commodity food aid to participating schools has just over $1 billion a year while NSLP cash assistance in FY2011 was over $10 billion.

In FY2011, more than 31.8 million children each day got their lunch through the National School Lunch Program.[165] Federal assistance for school food programs in FY2011 totaled approximately $14.4 billion, which consisted of $13.2 billion in cash assistance and $1.2 billion in donated food assistance.[166].

The National School Lunch Program accounted for the bulk of this assistance, with a total cost $11.1 billion in FY2011.[167]

Table 3. Annual USDA Food Commodity Purchases, Specialty Crops, FY2012 (YTD, ending August 10, 2012)

Commodity	Costs ($million)	Quantity (million pounds)
Apple	44.5	76.8
Apricot	5.3	7.0
Beans, Dried	20.0	43.6
Bean, Dry/Legume, Dry	11.2	16.0
Blueberry	1.7	1.2
Carrot	12.7	23.7
Cherry	2.3	2.7
Corn	10.7	19.2
Cranberry	5.1	9.6
Fruit Mix/Nut Mix	2.1	0.8
Grape Juice	9.6	13.3
Grapefruit Juice	0.5	1.1
Green Bean	15.6	32.2
Green Pea	8.0	13.2
Mixed Fruit	31.3	46.8
Mixed Vegetable	0.7	1.1
Orange & Orange Juice	10.6	20.0
Peach	46.3	65.8
Pear	26.3	41.0
Plum	15.7	12.0
Potato	28.1	101.3
Pumpkin	0.7	0.8
Raisin	3.8	2.4
Spinach	2.1	3.8
Strawberry	13.3	12.1
Sweet Potato	4.8	10.2
Tomato	33.4	84.2
Vegetable Soup/Chicken Soup	4.4	5.0
Subtotal, Fruits and Vegetables	370.6	666.9
Subtotal, Livestock and Seafood	316.9	148.1
Subtotal, Poultry Products	272.9	270.7
Total AMS Purchases To Date	$960.4	1,085.7

Source: AMS, "Annual Purchase Summary," http://www.ams.usda.gov/AMSv1.0/getfile?dDocName= STELPRDC5099583. FNS reports no purchases of almonds, asparagus, blackberry, date, fig, fruit cocktail, hominy, pistachio, raspberry, and walnut over the period. Historical data are at USD's website: http://www.ams.usda.gov/AMSv1.0/ams.fetchTemplateData.do?template=TemplateJ&page=CPD AnnualPurchaseSummary.

Table 4. Section 32 Contingency Fund (Bonus) Purchases, Specialty Crops, FY2000-FY2009

Commodity	Number of Years Purchased	Total Value Purchased (million $)	Commodity	Number of Years Purchased	Total Value Purchased (million $)
Almonds	3	29.5	Grapefruit	2	20.1
Apples	6	88.8	Mixed Fruit	2	79.5
Apricots	7	49.6	Orange Juice	5	99.5
Asparagus	7	28.3	Peaches	6	141.7
Beans	4	40.8	Pears	5	42.0
Blueberries	3	35.7	Pineapple	5	21.2
Caneberries[a]	3	4.5	Plums	4	8.2
Cherries	7	99.9	Potatoes[b]	6	113.2
Cranberries	5	80.6	Strawberries	3	12.8
Dates	3	7.2	Tomatoes	7	40.3
Figs	4	17.0	Trail Mix	4	78.5
Grape Products	6	95.0	Walnuts	6	94.8
			Specialty crops total		$1,246.7

Source: USDA and House Appropriations Committee, various hearing reports, supplemented by AMS unpublished data. Each category represents commodities and/or any foods processed from them, purchased by AMS. Purchases for each category are cumulative for the 10-year period covered. Does not include purchases of livestock, poultry, and seafood products. For other information, see CRS Report RL34081, Farm and Food Support Under USDA's Section 32 Program

[a] Includes raspberries and blackberries.

[b] Includes sweet potatoes.

The cash reimbursement to states is on the basis of the number of meals or snacks served to children at participating schools at reimbursement rates that vary according to household income status. USDA studies of school food purchase data, the relative share of school food purchases fresh fruits and vegetables (except for potatoes) appears to be increasing.[168] In recent years, Congress has substantially expanded support for fruit and vegetables within USDA's food and nutrition programs-both in the 2008 farm bill and in the 2010 reauthorization of child nutrition legislation (Healthy, Hunger-Free Kids Act, P.L. 111-296). The latter specifically required that USDA issue updated nutrition guidelines. The final regulation issued in January 2012 included requirements for participating schools to serve more fruits and vegetables.[169]

Health and nutrition concerns are likely to be among the leading drivers of increased demand for fresh fruits and vegetables in the next few years.[170]

Fresh Fruit and Vegetable ("Snack") Program

The Fresh Fruit and Vegetable Program (FFVP, also referred to as the Fruit and Vegetable "Snack" Program) gives cash grants to states and Indian reservations to provide free snacks of fresh fruits and vegetables to elementary school children during the school day.[171] FFVP started as a pilot program in the 2002 farm bill—funded with a one-time mandatory appropriation of $6 million—providing cash grants to selected states and Indian reservations, and covered both fresh and dried fruits and vegetables. FFVP was reauthorized and expanded in the 2004 Child Nutrition and WIC Reauthorization Act to include more states and reservations, also making it a permanent part of child nutrition law. The 2004 law provided mandatory funding of $9 million a year through FY2008. The 2008 farm bill permanently authorized the program nationwide and provided additional funding through Section 32, but limited purchases to fresh fruits and vegetables only. Funding for the program is permanent and now total more than $150 million per school year (taking into accounts necessary indexed cost adjustments). The program is administered through FNS. The agency expects to allocate to states a total of about $158 million for school year 2011/2012.[172] In most states, FFVP is primarily administered through states education agencies, except for Texas and New Jersey, where FFVP is administered by their agriculture agencies.

Assistance to Households and Families

A range of FNS programs provide foods for use in the home to individuals and families. These programs include the Commodity Supplemental Foods Program (CSFP), The Emergency Food Assistance Program (TEFAP), Food Distribution Programs on Indian Reservations (FDPIR), the Special Supplemental Nutrition Program for Women, Infants, and Children (WIC) and various disaster feeding programs. CSFP, WIC, and FDPIR provide specific foods based on the program's "food package" requirements.[173]

FNS's largest nutrition assistance program—based on participation and expenditures—is the Supplemental Nutrition Assistance Program (SNAP, formerly the Food Stamp Program), which provides benefits issued on Electronic Benefit Transfer (EBT) cards to households. SNAP benefits are not the same as cash: they are only redeemable at authorized stores, equipped with EBT machines, and may only be redeemed for SNAP-eligible foods. In general, SNAP benefits may be redeemed for any foods for home preparation and consumption, subject to certain exceptions.[174]

Previous studies by USDA indicate about 20% of SNAP benefits were spent on fruit and vegetable products (broadly defined) in FY2001.[175] Federal initiatives are being developed to further promote fruit and vegetable consumption under existing domestic nutrition assistance programs.[176]

Redeeming Nutrition Assistance Program Benefits at Farmers' Markets

Farmers' markets may accept EBT benefits and become SNAP-licensed retailers; they have done so at an increasing rate.[177] USDA reported that in 2010 that 1,611 farmers' markets or individual farmers were authorized to accept SNAP benefits, and they redeemed a total of $7.5 million in SNAP benefits—an increase of 263% in authorizations and a 49% in benefits redeemed compared to the previous five-year period.[178]

In addition to SNAP benefits, two other programs provide available resources to patronize and support farmers markets under two FNS-administered programs related to USDA's Farmer's Market Promotion Program (FMPP) (see "Farmer Direct Marketing Assistance"). These are:

- WIC Farmers' Market Nutrition Program (WIC-FMNP) and
- Senior Farmers' Market Nutrition Program (SFMNP).

These programs provide redeemable benefits to consumers at farmers' markets, allowing for farmers' market purchases (including fruit, vegetable, and tree nut products) by low-income WIC applicants and recipients and also low-income seniors, usually through the use of redeemable coupons.[179] Program benefits from SNAP and the farmers market nutrition programs may be redeemed for a range of agricultural products, including fruits, vegetables, and other specialty crops.

Certified Organic Foods

For some specialty crop producers, obtaining organic certification to sell their products as "organic" represents a viable business strategy. The Organic Foods Production Act (OFPA) of 1990 and USDA's National Organic Program (NOP) regulations require that agricultural products labeled as "organic" originate from farms or handling operations certified by a state or private entity that has been accredited by USDA.[180]

Organic agriculture accounts for a small but growing share of the U.S. farming sector. USDA reports that farm sales from organic fruit and vegetable operations totaled $1.2 billion in 2008, about 4% of all farm-level fruit and vegetable sales in the United States.[181] Following the 2008 farm bill, an average of approximately $54 million annually (FY2008-FY2012) in mandatory and discretionary program funding was authorized to be spent on certified organic agricultural production (*Table 2*). In addition to certified organic fruits, vegetables, and tree nuts, this annual estimate spans all certified organic production, including meat and dairy foods, as well as organic commodity crops. Among organic fruit and vegetable growers, USDA reports that there were approximately 3,900 vegetable farms, 3,300 fruit and tree nut farms, and 1,600 berry farms growing certified organic products. Ranked by acres in production, organic fruit and vegetable production is focused in California, Washington, Oregon, Florida, and Colorado.

Despite some shared program interests and also a shared title of the 2008 farm bill,[182] significant differences often exist between U.S. specialty crop and organic producers in terms of their overall farm bill priorities and in the types of key farm bill programs each group supports.

USDA programs supporting organic agricultural producers are spread across many different titles of the farm bill (see text box). The selected programs described in this report are administered by mostly by AMS and USDA's research and extension agencies. Other information about programs geared to certified organic producers is available in USDA's organic resource guide.[183]

Selected Organic Agriculture Provisions in the 2008 Farm Bill (P.L. 110-246)

Conservation (Title II)

- Organic Transition Support (Sec. 2501).
- Technical Asst. on Organic Conservation Practices (§ 2706)
- Organic Certification Crosslink (Sec. 2301).
- Organic Transition Incentives for Beginning Farmers (§ 2111)

Trade (Title III)

- Market Access Program (MAP) (§ 3102)

Credit (Title V)

- Organic Credit provision (§ 5002)

Research (Title VII)

- Organic Agriculture Research and Extension Initiative (§ 7206)
- Integrated Research, Education, and Extension Competitive Grants Program—Organic Transitions (ORG) (§7306)

Horticulture and Organic Agriculture (Title X)

- National Organic Certification Cost-Share Program (§ 10301)
- Organic Production and Marketing Data Collection (§ 10302)
- National Organic Program (§ 10303)

Crop Insurance and Disaster Assistance (Title XII)

- Organic Crop Insurance (Sec. 12023)

Source: Compiled by CRS.

National Organic Program

The National Organic Program (NOP), authorized by OFPA, is a regulatory program administered by AMS. The mission of the program is to "ensure the integrity" of USDA organic products, by overseeing the development of "national standards for organically-produced agricultural products to assure consumers that products with the USDA organic seal meet consistent, uniform standards."[184] The NOP regulations became operational in 2002, establishing a voluntary production and handling certification program that specifies the methods, practices and materials that may be used in how certified organic production is to be grown, raised, and processed.[185] Products labeled as "organic" must originate from farms or handling operations certified by a state or private entity that has been accredited by USDA.

Funding for the program covers regulatory enforcement and review and development of NOP regulations, among other activities such as responding to requests for international equivalency agreements. Funding is subject to appropriations, and annual authorizations have risen from under $2 million per year (FY2002-FY2007) to about \$7 million per year (FY2010-FY2012).[186] Authorized annual appropriations in the 2008 farm bill were \$11.0 million for FY2012 plus "such additional sums as are necessary."[187] No user fees are charged for appropriated activities.[188]

Organic Certification Cost-Share Program

Two USDA programs provide funding to reimburse eligible producers and handlers to offset the costs of NOP certification paid by producers to accredited agents for certification services. These programs include the National Organic Certification Cost-Share Program (NOCCSP)[189] and the Agricultural Management Assistance (AMA) Organic Certification Cost-Share Program.[190]

Both programs are administered by AMS. Detailed information about these two programs is available in USDA's annual reports to Congress and also USDA's annual fiscal year *Notice of Funds Availability* (NOFA) for each program.[191]

Both cost-share programs operate under a cooperative agreement between USDA and state agencies (typically state departments of agriculture). The state agencies process and review applications for cost share funds, which are reimbursed by USDA. Eligible applicants are producers and handlers who have received or renewed USDA organic certification from an accredited certifying agent. Applicants may be reimbursed for up to 75% of the cost of NOP certification, but no more than \$750 per year. The program is administered on a first-come, first-served basis until funds are exhausted.[192]

For FY2012, total available program funds for both cost-share programs was \$7.3 million.[193] Funds are made available each year on a state-by-state basis, with each state/territory receiving an allocation based on their historical activity and their number of certified organic operations. Each program has a separate funding allocation (and is separate from the NOP budget). Funding levels differ depending on the program. For NOCCSP, the 2008 farm bill allocated \$22.0 million on a one-time basis "to remain available until expended" from available mandatory CCC funds.[194] Accordingly, USDA obligated roughly \$5-\$6 million annually to states for reimbursements to farmers through FY2012.[195] For AMA funds, roughly \$1.5 million have been available annually over the same period through USDA appropriated funds.[196] Combined, both programs provided financial assistance to nearly 9,400 producers in FY2011.[197]

Product and Market Data Collection

USDA's Organic Production and Market Data Initiatives (ODI) builds on AMS *Market News* program (see section "Market News" in this report) and requires USDA to keep segregated data on the production and marketing of organic agricultural production, including price and market data. ODI was originally authorized in the 2002 farm bill, and expanded in the 2008 farm bill. The 2008 farm bill provided one-time funding of $5 million "to remain available until expended" from mandatory CCC funds, along with authorized appropriations of not more than $5 million annually (FY2008-FY2012) also "to remain available until expended."[198] USDA reports that it now covers about 246 different organic agricultural products—cotton, dairy and dairy products, fruits and vegetables, meat and grain, and poultry and eggs—while also developing additional organic market information tools within *Market News*.[199] ODI is administered by three USDA agencies: AMS, the Economic Research Service (ERS), and the National Agricultural Statistics Service (NASS).

Organic Agriculture Research and Extension Initiative

The Organic Agriculture Research and Extension Initiative (OREI) provides grants to facilitate the development of organic agriculture production, breeding, and processing methods through the integration of research and extension activities.[200] It funds projects intended to enhance the ability of producers and processors who have already adopted organic standards to grow and market high quality organic agricultural products. Priority concerns include projects addressing the biological, physical, and social sciences, including economics. A list of funded projects is available at USDA's website.[201] Funding is mandatory through the CCC: $18 million (FY2009) and $20 million annually (FY2010-FY2012), plus annual appropriations of $25 million (FY2009- FY2012). The program is administered by NIFA.

Organic Transitions Program

The Organic Transitions Program (ORG) funds research, extension, and education programs to improve the competitiveness of organic producers and those transitioning to organic practices, including the development and implementation of biologically based pest management practices.[202] ORG supports the development and implementation of research, extension and higher education programs to improve the competitiveness of organic livestock and crop producers, as well as producers who are adopting organic practices. A listing of funded projects is available at USDA's website.[203]

Appropriations of "such sums as necessary" are authorized through FY2012, and are currently estimated to total about $4.0 million annually.[204] The program is administered by NIFA.

EQIP Organic Initiative

As part of the Environmental Quality Incentives Program (EQIP), administered by USDA's Natural Resources Conservation Service (NRCS), the 2008 farm bill includes provisions to assist organic producers with natural resource concerns and requirements for the National Organic Program (NOP).[205] The EQIP Organic Initiative provides financial and technical assistance to implement approved conservation practices, and to develop and implement conservation plans (or Organic System Plans), and to assist producers who are transitioning to organic production. Eligible applicants include certified organic producers or producers: pursuing NOP certification, transitioning to organic production, or selling less than $5,000 organic products.[206] Assistance per producer is limited to $20,000 annually and $80,000 during a six-year period. USDA reports that the initiative funded contracts totaling $23.8 million (FY2010) and $22.5 million (FY2011) to producers across all states.207 In some states, obligations to producers totaled more than $1 million annually (California, Colorado, Iowa, Michigan, Minnesota, Missouri, Nebraska, New York, North Dakota, Washington, and Wisconsin).[208]

Other Farm Bill Programs

A number of other farm bill programs assist specialty crop producers that are not specifically addressed in this report. Many of these programs are important to some specialty crop producers. For example, in the commodities title, the 2008 farm bill authorized a pilot project in selected states[209] to allow fruits and vegetables grown on acreage participating in the commodity support programs to be used for processing on up to 75,000 acres. Expansion of "planting flexibility" policies that would further allow growers who receive federal payments to also plant fruits and vegetables on acres on which they receive benefits (base acres) is generally opposed by the major specialty crop groups.[210]

Another example is conservation programs that may benefit all eligible U.S. agricultural producers, including specialty crop growers. Changes in recent farm bills expanded incentives to encourage greater farmer participation through cost-sharing and technical assistance programs, such as Environmental Quality Incentives Program (EQIP), Conservation Stewardship Program (CSP), and Conservation Reserve Program (CRP), as well as competitive grants including Conservation Innovation Grants, Cooperative Conservation Partnership Initiatives, and Conservation Technical Assistance.[211]

Labor Protections

Most fruit and vegetable production, processing and distribution—from planting and harvesting to packaging and transportation—is highly labor-intensive, making produce growers especially dependent on hired and contract labor. USDA reports that labor expenses at fruit and vegetable farms total 48% and 36%, respectively, as a share of total variable expenses.[212] This compares to 17% across all farms, and 5%-6% on corn and soybean farms. Reportedly, the fruit and vegetable industry often relies upon a mostly-immigrant workforce and faces labor shortages in some produce growing areas.[213]

H-2A Program

Under current law, certain lower-skilled foreign workers, sometimes referred to as guest workers, may be admitted to the United States to perform temporary service or labor under temporary worker visas. For agricultural workers, the H-2A visa program[214] establishes a means for agricultural employers who anticipate a shortage of domestic workers to bring nonimmigrant foreign workers to the U.S. to perform agricultural labor or services of a temporary or seasonal nature. It allows employers to petition for the temporary admission of foreign workers to the United States to perform agricultural labor or services of a seasonal or temporary nature, provided that U.S. workers are not available.[215] Eligible applicants under the H-2A program are agricultural employers and may include an individual proprietorship, a partnership or corporation, or an association of agricultural producers.

The program is administered by the Department of Labor (DOL), the Department of Homeland Security (DHS), and the Department of State. Employers must demonstrate to DOL that sufficient domestic workers are not available and that employment of foreign workers will not adversely affect U.S. workers who are similarly employed. DHS handles the visa determinations. After receiving a labor certification from DOL, an employer petitions DHS for approval to hire foreign workers. A Department of State foreign office issues the visas.[216]

Farmworker Assistance Programs

DOL administers a number of programs intended to benefit domestic agricultural workers, whose lives tend to be characterized by poverty, frequent moving, and chronic unemployment and underemployment. One such program is DOL's Employment and Training Administration (ETA), the National Farmworker Jobs Program (NFJP).[217] NFPJ provides grants for services provided by state/local government agencies and private non-profit institutions and organizations that operate employment and training programs. Grantees provide job training and other employment and education services and related assistance to migrant and seasonal farmworkers (MSFWs) to address chronic seasonal unemployment and underemployment, and to increase the income and stability of farmworker families. DOL also administers and enforces requirements under the Migrant and Seasonal Agricultural Worker Protection Act,[218] which provides for certain employment-related protections to migrant and seasonal agricultural workers, and is responsible for monitoring farm labor contractors and the wages, working conditions, and housing arrangements of migrant and seasonal laborers, among other things. DOL's Occupational Safety and Health Administration administers workplace and field safety and sanitation requirements. Additionally, there are special provisions for the education of farmworkers' children under the No Child Left Behind Act (Department of Education).[219]

End Notes

[1] Defined in the Specialty Crops Competitiveness Act of 2004 (P.L. 108-465, Section 3), as amended by the 2008 farm bill (P.L. 110-246, Food, Conservation, and Energy Act of 2008). See USDA, "USDA Definition of Specialty Crop," http://www.ams.usda.gov/AMSv 1.0/getfile?dDocName=STELPRDC5082113. Excludes peanuts, other commodities.

[2] USDA, NASS, 2007 Census of Agriculture, Specialty Crops, vol. 2, November 2009. USDA reports 13.766 million acres in specialty crops in 2007. Compares to total cropland acres of 406.4 million acres in 2007.

[3] As reported by USDA, AMS in "PACA—Your Partner in Produce." Reflects estimates for 2008.

[4] Compiled by CRS from data in the U.S. International Trade Commission's Trade DataWeb database. Includes fresh and processed fruits, vegetables, and tree nuts (excluding peanuts), and live trees and plants.

[5] USDA, NASS, 2007 Census of Agriculture, Specialty Crops, vol. 2, November 2009. Most recent data available.

[6] See, for example, USDA, Production Expenses of Specialized Vegetable and Melons Farms, VGS-328-01, September 2008, and Specialized U.S. Fruit and Nut Farm Production Expenses, FTS-337-01, June 2009. Data vary by region.

[7] "Commodity crops" refers to agricultural production supported by specific federal farm support programs (mostly corn, soybeans, wheat, cotton, rice, dairy, and sugar).

[8] See CRS Report R41195, Actual Farm Bill Spending and Cost Estimates, Table 1.

[9] See, e.g., SCFBA, "2012 Farm Policy Recommendations," http://www.strongeragriculture.org/#/policy-priorities.

[10] Food, Conservation, and Energy Act of 2008, P.L. 110-246.

[11] Farm Security and Rural Investment Act of 2002, P.L. 107-171.

[12] CFDA has detailed program descriptions for more than 2,000 federal assistance programs (https://www.cfda.gov).

[13] 76 Federal Register 37312, June 27, 2011. See also USDA, http://www.ams.usda.gov.AMSv1.0/fv.

[14] Federal Crop Insurance Act, as amended, 7 U.S.C. § 1501-1520 (CFDA# 10.450).

[15] For more information, see CRS Report R40532, Federal Crop Insurance: Background and Issues; and CRS Report RS21212, Agricultural Disaster Assistance.

[16] In other words, it excludes crops where federal crop insurance coverage is not provided; it also excludes nursery and floriculture products that are also considered "specialty crops" per the statutory definition.

[17] Calculated as 7.25 million insured acres divided by 12.3 million total acres in fruits, vegetables, and tree nuts (USDA's 2007 Census of Agriculture, Specialty Crops, excluding acreage for nursery and floriculture crops which are not included in the RMA figure for insured acres).

[18] USDA, Federal Crop Insurance Corporation (FCIC), Report to Congress: Specialty Crop Report, November 2010, Appendix A (Tables 1 and 2), http://www.rma.usda.gov/pubs/2010/specialtycrop.pdf. Actual estimates may vary depending on how different crop varieties are counted. See also CRS Congressional Distribution Memo, "Crop Insurance for Specialty and Organic Crops," December 2, 2011.

[19] Keith Collins, "Crop Insurance and Specialty Crops," Crop Insurance Today, August 2012.

[20] Ibid. USDA, FCIC, Report to Congress: Specialty Crop Report, November 2010.

[21] USDA, FCIC, Report to Congress: Specialty Crop Report, November 2010.

[22] Annual USDA appropriations acts provide funding for RMA salaries and expenses to operate the program. The crop insurance program receives such sums as are necessary for premium subsidy and program losses and expenses, which makes it a mandatory program.

[23] Keith Collins, "Crop Insurance and Specialty Crops," Crop Insurance Today, August 2012.

[24] Keith Collins, National Crop Insurance Services, "Crop Insurance for Specialty Crops." Presentation to the Specialty Crop Caucus, House of Representatives, April 12, 2012.

[25] USDA, "Noninsured Crop Disaster Assistance Program (NAP) for 2011 and Subsequent Years," August 2011, http://www.fsa.usda.gov/Internet/FSA_File/nap_august_2011.pdf. The regulatory definition of a NAP-eligible crop is one for which catastrophic coverage is not available and which is commercially produced for food or fiber as specified in the regulations. The term also includes floriculture, ornamental nursery, Christmas tree crops, turfgrass sod, seed crops, aquaculture (including ornamental fish), and industrial crops.

[26] As amended by the Federal Agriculture Improvement and Reform Act of 1996 (P.L. 104-127). CFDA# 10.451.

[27] USDA's CCC is a government-owned corporation that is authorized to borrow up to $30 billion at any one time from the U.S. Treasury. The CCC mainly is a financing mechanism for farm bill programs such as commodity price and income supports, agricultural conservation, export assistance, and other mandated authorizations.

[28] For more information, see CRS Report R40532, Federal Crop Insurance: Background and Issues; and CRS Report RS21212, Agricultural Disaster Assistance.

[29] FAS, "Disaster Assistance Programs," http://www.fsa.usda.gov/FSA/webapp? area= home&subject=diap&topic=tap.

[30] P.L. 110-246, § 12033; 7 U.S.C § 1501 et seq. (CFDA# 10.092). USDA, "Tree Assistance Program for Orchardists and Nursery Tree Growers (TAP)," http://www.fsa.usda.gov/ Internet/FSA_File/tap051010.pdf.

[31] Trade Act of 1974 (P.L. 93-618), as amended; 19 U.S.C § 2947a (CFDA# 10.091). USDA, "Emergency Assistance for Livestock, Honeybees, and Farm-Raised Fish Program (ELAP)," http://www.fsa.usda.gov/Internet/FSA_File/ elap_livestock_2011.pdf.

[32] P.L. 110-246, § 12033; 7 U.S.C § 1501 et seq. USDA, http://www.fsa.usda.gov/ Internet/FSA_File/sure_2011.pdf.

[33] Consolidated Farm and Rural Development Act, as amended; 7 U.S.C. § 1961- 1984 (CFDA# 10.404). USDA, http://www.fsa.usda.gov/FSA/webapp? area=home&subject=fmlp &topic= efl.

[34] P.L. 110-246, § 10404. Also USDA, http://www.fsa.usda.gov/FSA/webapp?area=home& subject=prsu&topic=mpp. USDA-administered market loss programs generally compensate agricultural producers for specific market disruptions.

[35] 7 U.S.C. § 499a et seq., and § 1622, respectively (CFDA#10.165). Regulations are at 7 CFR Part 46.

[36] USDA, AMS presentation, "PACA—Your Partner in Produce." See also USDA, http://www.ams.usda.gov/paca. Exemptions include growers who handle only their own product and truckers who haul for hire only.

[37] Ibid.

[38] USDA, "2013 Explanatory Notes, AMS," http://www.obpa.usda.gov/19ams2013notes.pdf, p. 19-67.

[39] P.L. 108-465, as amended; 7 U.S.C. § 1621 note (CFDA# 10.170). Also see USDA, http://www.ams.usda.gov/scbgp.

[40] The minimum base grant each state is eligible to receive is equal to the higher of $100,000 or 1/3 of 1% of the total amount of funding made available for that fiscal year. For FY2010, the base grant portion was $181,210 per state. The additional allocation is based on the value of specialty crop production in each state relative to national production, using available cash receipt data.

[41] USDA, "Fiscal Year 2011 Awards," http://www.ams.usda.gov/AMSv 1.0/getfile? dDocName=STELPRDC5093883.

[42] Including Arizona, Georgia, Idaho, Michigan, New York, North Carolina, Oregon, Pennsylvania, Texas, Wisconsin.

[43] AMS, "Funded Projects," http://www.ams.usda.gov/AMSv 1.0/getfile?dDocName= STELPRDC5093992.

[44] Ibid. AMS' report provides a full listing of all program recipients by state, applicant name, and grant amount.

[45] P.L. 106-224, § 6202, as amended; 7 U.S.C. § 1621 note (CFDA# 10.352).

[46] USDA, http://www.rurde.usda.gov/rbs/coops/vadg.htm.

[47] However, USDA's FY2013 budget justification indicates that annual program funding has remained under $15 million per year (see USDA, 2013 Explanatory Notes, Rural Business-Cooperative Service," http://www.obpa.usda.gov/28rbs2013notes.pdf, p. 73).

[48] USDA, http://www.rurdev.usda.gov/supportdocuments/BCP_FY2011VAPGAWARDS.pdf (recipients by state, applicant name, and grant amount). Awards for previous years (2001-2010) are also available at USDA's website.

[49] P.L. 94-463, as amended; 7 U.S.C. § 3005 (CFDA# 10.168). AMS, http://www.ams.usda.gov/AMSv1.0/FMPP.

[50] USDA, "Grants, Loans, and Support," http://www.usda.gov/wps/portal/usda/usdahome?navid=KYF_GRANTS.

[51] FNS, "Grant Levels by State, FY 2006-2011," http://www.fns.usda.gov/ wic/ FMNP/ FMNPgrantlevels.htm.

[52] FNS, "SFMNP Grant Levels, FY 2006-2011," http://www.fns.usda.gov/wic/SeniorFMNP/SFMNPgrantlevels.htm.

[53] For other information see CRS Report R42155, The Role of Local Food Systems in U.S. Farm Policy.

[54] 7 U.S.C. §§ 1621 et seq.; CFDA# 10.153. USDA, "2013 Explanatory Notes, AMS," http://www.obpa.usda.gov/ 19ams2013notes.pdf, p. 19-1.

[55] Other commodities are cotton, cottonseed, tobacco, dairy products, livestock, meat, grains, wool, poultry and eggs.

[56] P.L. 110-246, § 10107. 7 U.S.C. § 1622b(b).

[57] Marketing orders and research and promotion programs for certain fruit and vegetable crops have come under legal challenge from producers. See CRS Report 95-353, Federal Farm Promotion ("Check-Off") Programs.

[58] 7 U.S.C. § 601 et seq. (CFDA# 10.155).

[59] USDA, "Marketing Orders and Agreements," http://www.ams.usda.gov/fv/.

[60] 76 Federal Register 24292, April 29, 2011.

[61] USDA, "Proposed NLGMA, Q&A" http://www.ams.usda.gov/AMSv 1.0/getfile?dDocName=STELPRDC5090539.

[62] For other background information, see LGMA's website (http://www.caleafygreens.ca.gov/).

[63] See, for example, National Farmers Union (NFU), "Proposed National Leafy Green Marketing Agreement is the Wrong Direction for Food Safety," June 21, 2011, http://www.nfu.org/blog/?p=48; and Center for Science in the Public Interest (CSPI), Comments submittedtoUSDA on the proposed NLGMA, July 28, 2011, http://www.cspinet.org/foodsafety/PDFs/CommentOnNatlLeafyGreensMktAgreement-072811.pdf.

[64] See, for example, National Sustainable Agriculture Coalition, "Opposition to NLGMA Rises," June 8, 2011, http://sustainableagriculture.net/blog/opposition-to-nlgma-rises/.

[65] 7 U.S.C. §§ 1621 et seq. (CFDA# 10.162). Regulations are at 7 CFR Part 75.

[66] All grading services in Oklahoma are currently performed by AMS.

[67] USDA, "2013 Explanatory Notes, AMS," http://www.obpa.usda.gov/19ams2013notes.pdf, p. 19-45.

[68] USDA, "USDA Quality Standards," http://www.ams.usda.gov/standards/.

[69] 7 U.S.C. §§ 1621-1627, 7 U.S.C. §§ 2101-2119, as amended (CFDA# 10.163).

[70] FFDCA § 401; 21 U.S.C. § 341. Regulations are 21 CFR Parts 130-169. 21 CFR 130 covers general requirements.

71 Canned fruits (Part 145); fruit juices and beverages (Part 146); jellies and preserves (Part 150); fruit pies (Part 152); canned vegetables (Part 155); vegetable juices (Part 156); frozen vegetables (Part 158); tree nut products (Part 164).

[72] See, for example, FDA, "FDA's Standards for High Quality Foods," http://www.fda.gov/For Consumers/Consumer Updates/ucm094559.htm. Also see North Dakota State University, "Standard of Identity, Food Additives and Claims," http://www.ag.ndsu.edu/foodlaw/ processingsector/standardofidentity.

[73] Procedures for establishing food standards are at 21 CFR 130.5 FDA's guidance for industry is at http://www.fda.gov/downloads/Food/GuidanceComplianceRegulatoryInformation/ GuidanceDocuments/ FoodIngredientsandPackaging/RegulatorySubmissions/UCM201599. pdf. A listing of existing international standards for foods is available from the Codex Alimentarius website: http://www.codexalimentarius.org/standards/list-ofstandards/en/.

[74] P.L. 107-171, § 10816, amending the Agricultural Marketing Act of 1946; 7 U.S.C. §§ 1638 et seq. Regulations for covered commodities are at 7 CFR Part 60 and Part 65.

[75] Specialty crops per USDA, "USDA Definition of Specialty Crop," http://www.ams. usda.gov/ AMSv1.0/getfile? dDocName=STELPRDC5082113.

[76] 74 Federal Register 2658-2707, January 15, 2009. Also see USDA, Office of Inspector General "Implementation of Country of Origin Labeling," August 2011, http://www. usda.gov/oig/webdocs/01601-04-HY.pdf

[77] 74 Federal Register 2658-2707, January 15, 2009. Table 1.

[78] United Fresh Produce Association (UFPA), "Country of Origin Labeling," http://www.united fresh.org/newsviews/ country_of_origin_labeling. Other information is at Produce Marketing Association, "Country of Origin Labeling," http://www.pma.com/resources/ issues-monitoring/country-origin-labeling.

[79] CSPI, Outbreak Alert! Database search for "Produce" (http://www.cspinet.org/foodsafety/ outbreak/pathogen.php#). Produce includes fruits, vegetables, and dishes containing fruits or vegetables. CSPI's database includes outbreaks where both the food and pathogen have been identified and currently has information on over 6,000 outbreaks that occurred between 1990 to 2009.

[80] FDA, "Guidance for Industry Guide to Minimize Microbial Food Safety Hazards for Fresh Fruits and Vegetables," October 1998, http://www.fda.gov/downloads/Food/Guidance ComplianceRegulatoryInformation/GuidanceDocuments/ ProduceandPlanProducts/UCM169112.pdf.

[81] For more information, see CRS Report RS22600, The Federal Food Safety System: A Primer.

[82] P.L. 111-353; 21 U.S.C. §§ 301 et seq.

[83] FSMA explicitly exempts certain food processors and farms from FSMA if they are either a "very small business" as defined by FDA in rulemaking, or if the facility's or farm's "average annual monetary value" of all food sold during the previous three year period was less than $500,000, provided that the food is sold directly to certain "qualified end users" located in the same state where the facility or farm sold the food or within 275 miles of the facility or farm.

[84] These are the types of production areas identified in FDA's 1998 guidance (FDA, "Guidance for Industry Guide to Minimize Microbial Food Safety Hazards for Fresh Fruits and Vegetables," October 1998).

[85] See, for example, FDA, "Guidance for Industry, Necessity of the Use of Food Categories in Food Facility Registrations and Updates to Food Categories," August 2012, draft submitted for public comment.

[86] The Public Health Security and Bioterrorism Preparedness and Response Act of 2002 (P.L. 107-188; 7 U.S.C. 8401).

[87] 76 Federal Register 24292, April 29, 2011.

[88] For more information, see AMS, "Qualified Through Verification"(QTV) Program for the Fresh-Cut Produce Industry," July 2012, http://www.ams.usda.gov/AMSv1.0/ getfile?dDocName=stelprdc5059800. HACCP refers to a management system that addresses food safety through the analysis and control of biological, chemical, and physical hazards from raw material production, procurement and handling, to manufacturing, distribution and consumption of the finished product.

[89] Congress provided AMS with funds to initiate MDP in the FY2001 agriculture appropriations bill (amending the Agricultural Marketing Agreement Act of 1946).

[90] See USDA's website (http://www.ams.usda.gov/AMSv1.0/mdp). For other information, see USDA's list of produce commodities tested between 2001 (http://www.ams.usda.gov/ AMSv1.0/getfile?dDocName=STELPRDC5059955) and also AMS, "Microbiological Data Program Progress Update and 2009 Data Summary," January 2011, http://www.ams. usda.gov/AMSv1.0/getfile?dDocName=STELPRDC5088761.

[91] D. Acheson, "Should the Microbiological Data Program be Saved?" Food Safety News, July 27, 2012.

[92] USDA, "2013 Explanatory Notes, AMS," http://www.obpa.usda.gov/19ams2013notes.pdf, p. 19-89.

[93] For information on U.S. pesticide laws, see CRS Report RL31921, Pesticide Law: A Summary of the Statutes.

[94] Among some of the foods surveyed are canned black beans, orange juice, apples, grapes, pears, asparagus, hot peppers, sweet bell peppers, fresh and frozen sweet corn, green beans, canned spinach, cabbage, pears, fresh and canned spinach, cilantro, sweet potatoes, cantaloupe, lettuce, watermelon, mangoes, canned garbanzo beans, cucumbers, and oranges, and also green beans, pears, and sweet potatoes used in baby food.

[95] USDA, "Quick Facts about the Pesticide Data Program (PDP)," http://www.ams. usda.gov/AMSv1.0/getfile? dDocName=STELDEV3003252.

[96] AMS, Pesticide Data Program: Annual Summary Calendar Year 2010, May 2010, http://www.ams.usda.gov/ AMSv1.0/getfile?dDocName=stelprdc5098550.

[97] P.L. 95-501, as amended; 7 U.S.C. § 5623 (CFDA# 10.601). Regulations are at 7 CFR Part 1485. Also see USDA, http://www.fas.usda.gov/mos/programs/map.asp.

[98] See USDA, "Market Access program (MAP)," http://www.fas.usda.gov/mos/ programs/map.asp. Branded promotion requires a 50% match whereas generic promotion requires a 10% match.

[99] P.L. 110-246, § 3102.

[100] This amount does not include support through other national, state, or regional export promotion groups that might also provide support for specialty crops, among other agricultural commodities. USDA, FAS, "Fiscal Year 2010 Market Access Program Allocations," http://www.fas.usda.gov/info/factsheets/MAP.asp.

[101] Section 5(f) of the CCC Charter Act, as amended; 15 U.S.C. 714c(f) (CFDA# 10.605). USDA, http://www.fas.usda.gov/mos/programs/qsp.asp.

[102] Agricultural Trade Act of 1978, as amended; 7 CFR 1493, Part A (CDFA# 10.610). USDA, http://www.fas.usda.gov/excredits/ecgp.asp.

[103] FAS, "Eligible Commodities Under the GSM-102 Program," http://www.fas.usda.gov/ excredits/ gsmcommodities.html.

[104] Agricultural Trade Act of 1978, Title VII, 7 U.S.C. § 5721, et seq. (CDFA# 10.600).

[105] USDA, "Foreign Market Development Program," http://www.fas.usda.gov/info/factsheets/ fmd.asp. Funding for FMD was $34.2 million (FY2010), of which no funds were directed toward specialty crop groups.

[106] Food for Progress Act of 1985, as amended; 7 U.S.C. § 1736o (CFDA# 10.606). USDA, http://www.fas.usda.gov/excredits/foodaid/ffp/foodforprogress.asp.

[107] Food aid data (2009) by commodity (http://www.fas.usda.gov/excredits/ FoodAid/Reports/ 2009FoodAidTable3.pdf).

[108] P.L. 107-171, § 3205, as amended; 7 U.S.C. 5680 (CFDA#10.604); additional authorization were made in the Specialty Crop Competitiveness Act of 2004 (P.L. 108-465). Regulations are at 7 CFR part 1487. FAS, "Technical Assistance for Specialty Crops Program," http://www.fas.usda.gov/info/factsheets/Final_TASC_January%202011.pdf.

[109] The so-called SPS Agreement entered into force on January 1, 1995, as part of the establishment of the WTO, following the Uruguay Round of the General Agreement on Tariffs and Trade (GATT). The TBT Agreement resulted from the Tokyo Round in 1979. For more information, see CRS Report RL33472, Sanitary and Phytosanitary (SPS) Concerns in Agricultural Trade.

[110] USTR, 2011 Report on Sanitary and Phytosanitary Measures and 2011 Report on Technical Barriers to Trade, March 2011, available at USTR's website (http://www.ustr.gov). For other information on how SPS barriers may affect specialty crop trade, see CRS Report RL34468, The U.S. Trade Situation for Fruit and Vegetable Products.

[111] Trade Act of 1974, as amended by the Trade Act of 2002 (P.L. 107-210), and the 2009 economic stimulus package (P.L. 111-5) and later in 2010 (P.L. 112-40). 19 U.S.C. § 2401 (CFDA# 10.609). Regulations are at 7 C.F.R. § 1400.3. FAS (http://www.fas. usda.gov/itp/taa/). See also CRS Report R40206, Trade Adjustment Assistance for Farmers.

[112] Covers producers of raw and natural agricultural commodities (crops, livestock, farm-raised aquatic products, and wild-caught seafood that competes with aquaculture products).

[113] 19 U.S.C. § 1673 et seq. and 19 U.S.C. 1671 et seq. Regulations are at 19 C.F.R. Parts 207 and 351. Other information is in CRS Report RL32371, Trade Remedies: A Primer.

114 USITC, Conditions of Competition for Certain Oranges and Lemons in the U.S. Fresh Market, Inv. 332-469, July 2006; USITC, Canned Peaches, Pears, and Fruit Mixtures: Conditions of Competition between U.S. and Principal Foreign Supplier Industries, Inv.332-485, December 2007. Reports available at http://www.usitc.gov.

[115] USITC, "Import Injury Investigations Case Statistics (FY 1980-2009)," February 2010, http://www.usitc.gov/ trade_remedy/documents/historical_case_stats.pdf. Other information is in CRS Report RL34468, The U.S. Trade Situation for Fruit and Vegetable Products.

[116] For more information, see APHIS, "A 40-Year Retrospective of APHIS, 1972–2012."

[117] 7 U.S.C. § 7701 et seq. PPA became law in 2000 as part of the Agricultural Risk Protection Act (P.L. 106-224) and consolidated and superseded several U.S. plant health laws, including (1) The Act of August 20, 1912 (7 U.S.C. § 151- 164a, 167); (2) The Federal Plant Pest Act (7 U.S.C. § 150aa et seq. and 7 U.S.C. 147a); (3) Section 102 (a) - (e) of the Department of Agriculture Organic Act of 1944 (7 U.S.C. § 147a); (4) The Federal Noxious Weed Act of 1974 (7 U.S.C. § 2801 et seq.), except §§ 1-15 (7 U.S.C. § 2801 note and 7 U.S.C. § 2814); (5) Joint Resolution of April 6, 1937 (7 U.S.C. § 148 et seq.); (6) The Halogeton Glomeratus Act (7 U.S.C. § 1651 et seq.); (7) The Golden Nematode Act (7 U.S.C. § 150 et seq.); and (8) Section 1773 of the Food Security Act of 1985 (P.L. 99-198; 7 U.S.C. § 148f).

[118] PPA (7 U.S.C. §§ 7751, 7772, 431 and 442); also AHPA (7 U.S.C. §§ 8310, 8316, 10411 and 10417), replacing previous authorities in other laws. For more information, see CRS Report RL32504, Funding Plant and Animal Health Emergencies: Transfers from the Commodity Credit Corporation.

[119] USDA, "2013 Explanatory Notes, APHIS," http://www.obpa.usda.gov/18aphis 2013notes.pdf, p. 18-108.

[120] P.L. 110-246, § 10201; 7 U.S.C. § 7721. USDA's website (http://www.aphis.usda.gov/plant_health/plant_pest_info/ pest_detection/farm_bill.shtml).

[121] P.L. 110-246, § 10202; 7 U.S.C. § 7761. USDA (http://www.aphis.usda.gov/plant_health/ncpn/index.shtml); USDA's fact sheet (http://www.aphis.usda.gov/plant_health/ncpn/downloads/NCPN-FactSheet2010-11-22.pdf).

[122] See, for example, APHIS, "White Paper, Section 10201(d)(1) of the Farm Bill pertaining to Audit-based Certification," http://www.nationalplantboard.org/docs/sanc_Farm_ Bill_10201d_White_Paper_draft_1.pdf.

[123] USDA, "2013 Explanatory Notes, APHIS," http://www.obpa.usda.gov/18aphis2013 notes.pdf, p. 18-104-106.

[124] These include either universities or state agencies in Arkansas, Alabama, Arizona, California, Florida, Hawaii, Louisiana, Michigan, Missouri, New York, North Carolina, Oregon, South Carolina, Texas, and Washington.

[125] Comments by Rayne Pegg, AMS Administrator, before the House Subcommittee on Horticulture and Organic Agriculture, October 28, 2009. Amends 36 CFR 230 (see 77 Federal Register 7900, February 13, 2012).

[126] USDA, "2013 Explanatory Notes, APHIS," http://www.obpa.usda.gov/18aphis2013notes. pdf, p. 18-15.

[127] APHIS' website (http://www.aphis.usda.gov/plant_health/plant_pest_info/pest_ detection/ index.shtml). Also see APHIS, "2008 Farm Bill: Plant Pest and Disease Management and Disaster Prevention," July 2012.

[128] USDA, "2013 Explanatory Notes, APHIS," http://www.obpa.usda.gov/ 18aphis 2013notes.pdf, p. 18-15.

[129] Until 2002, APHIS held sole responsibility for operating the AQI. In 2002, in the law creating DHS (P.L. 107-296), Congress transferred the inspection function and more than 2,600 APHIS inspectors to DHS.

[130] Quarantine regulations are at 7 CFR part 319 and apply to many commodities, including nursery stock.

[131] For more information, see APHIS' website (http://www.aphis.usda.gov/plant_health/) and also "Plant Inspection Station Strategic Plan, 2007-2012."

[132] USDA, "2013 Explanatory Notes, APHIS," http://www.obpa.usda.gov/ 18aphis2013notes. pdf, p. 18-15 and 18-30.

[133] See USDA's website (http://www.aphis.usda.gov/plant_health/cphst/npgqbl.shtml).

[134] The FAVIR database is available at https://epermits.aphis.usda.gov/manual/index. cfm?ACTION=pubHome.

[135] See USDA's website: http://www.aphis.usda.gov/plant_health/plant_pest_info/index.shtml.

[136] For more information, see APHIS' website: http://www.aphis.usda.gov/import_export/sanitary_phytosanitary.shtml.

[137] USDA, "2013 Explanatory Notes, APHIS," http://www.obpa.usda.gov/18aphis 2013 notes.pdf, p. 18-58.

[138] See USDA's FAQs (http://www.aphis.usda.gov/import_export/plants/ plant_ exports /faqs.shtml) and other information at http://www.aphis.usda.gov/import_export/plants/ plant_exports/export_certificates_forms.shtml.

[139] Formerly Cooperative State Research, Education, and Extension Service or CSREES.

[140] For information, see USDA (http://www.agcensus.usda.gov/Publications/ Census_ of_Horticulture_Specialties/).

[141] NIFA, "Cooperative Extension Offices," http://www.csrees.usda.gov/Extension/; map (http://www.csrees.usda.gov/qlinks/partners/partners_map.pdf).

[142] Appropriate Technology Transfer for Rural Areas (ATTRA) project, authorized in the 1985 farm bill. For more information, see CRS Report RL31837, An Overview of USDA Rural Development Programs.

[143] P.L. 101-246, § 7311 (amended the 1998 Agricultural Research, Extension, and Education Reform Act, AREERA); 7 U.S.C. § 7632 et seq. (CFDA# 10.309). In establishing the initiative, the 2008 farm bill also removed specialty crop research from USDA's list high priority research and extension activities (7 U.S.C. § 5925), which had been added in the 2005 Specialty Crop Competitiveness Act (P.L. 106-465, § 302).

[144] USDA's website (http://www.csrees.usda.gov/fo/specialtycropresearchinitiative.cfm.) under "Abstracts of Funded Projects" (recipient names, award amount, and project terms).

[145] The program followed the 1995 formation of the Methyl Bromide Alternatives Working Group to track and facilitate adoption of alternatives to methyl bromide, and the allocation of ARS funds starting in FY1999 toward research to develop alternatives to methyl bromide, principally at the University of California and the California Strawberry Commission, and the University of Florida. ARS, "National Program 308: Methyl Bromide Alternatives Strategic Vision," http://www.ars.usda.gov/research/programs/ programs.htm?np_code=308.

[146] EPA, "The Phaseout of Methyl Bromide," http://www.epa.gov/ozone/mbr/.

[147] CUEs are permitted under Section 604(d) of the Clean Air Act and also under the Protocol.

[148] Squashes, melons, cucumbers, and gourds, etc.

[149] EPA, "List of Critical Uses," http://www.epa.gov/ozone/mbr/cueuses.html; and USDA, "National Program 308 MeBr Alternatives," http://www.ars.usda.gov/research/programs/ programs.htm?np_code=308&docid=14368.

[150] United Nations Environmental "Report of the Technology and Economic Assessment Panel," May 2011, http://ozone.unep.org/teap/Reports/ TEAP_Reports/TEAP_ Progress_ Report_May_2011.pdf, p. 149.

[151] This report does not cover spending on fruit, vegetable, and tree nut products financed under nutrition programs authorized by the Older Americans Act (administered by the Department of Health and Human Services), for which no information regarding specific food types of food purchases is available, nor does it address federally supported nutrition education initiatives aimed at increasing consumption of fruits and vegetables.

[152] In this case, the term "commodity foods" refers to all USDA purchased foods, which includes fruits and vegetables, and livestock, poultry, and seafood products, and applies more broadly than "commodity crops."

[153] USDA FNS, "White paper: USDA Foods in the National School Lunch Program," May 2010, http://www.fns.usda.gov/fdd/foods/healthy/WhitePaper.pdf. AMS generally serves as the commodity purchasing agency, and the Farm Service Agency also assists in making commodities available.

[154] For more information, see CRS Report RL34081, Farm and Food Support Under USDA's Section 32 Program.

[155] USDA FNS, "Food Distribution Programs Overview," September 2011. For information on these and other domestic assistance programs, see CRS Report R42353, Domestic Food Assistance: Summary of Programs.

[156] NSIP is jointly administered by HHS and USDA's FNS.

[157] 7 U.S.C. § 612c. Section 32 requires that 30% of annual customs receipts be used by USDA to buy U.S. agricultural commodities. For information see CRS Report RL34081, Farm and Food Support Under USDA's Section 32 Program.

[158] FNS, Purchases by Product, USDA Fruit and Vegetable Programs, Fiscal Year 2010 (10/01/09-09/30/10), http://www.ams.usda.gov/AMSv1.0/getfile?dDocName= STELPRDC 5088443. Historical data are available at AMS, "Commodity Purchasing," http://www.ams. usda.gov/AMSv1.0/ams.fetchTemplateData.do?template= TemplateJ&page=CPDAnnual PurchaseSummary. Total purchases exceeded roughly $1 billion in FY2010.

[159] The 2002 farm bill provided $200 million annually and each year thereafter (P.L. 107-171 § 10603); the 2008 farm bill provided an additional $206 million for FY2012 and each year thereafter (P.L. 110-246, § 4404).

[160] DoD Fresh is a mechanism created by USDA to increase fresh produce offerings to schools. DoD Fresh, which utilizes the logistical capacity of the United States military to delivery food to U.S. military bases across the country and world, began as a USDA pilot project in 1996. This program now operates in more than 40 states. The program works in partnership with USDA to take advantage of DoD's buying power, distribution system, and nationwide network of suppliers.

[161] Entitlement funding for USDA foods totaled $1.057 billion; bonus funding totaled $0.178 billion.

[162] FNS, "USDA Foods," http://www.fns.usda.gov/fdd/foods/healthy/CNP_PPT__LONG-FINAL.pdf, p. 8.

[163] 31 U.S.C. § 1535.

[164] For more information, see CRS Report R42353, Domestic Food Assistance: Summary of Programs.

[165] FNS, "National School Lunch Program," http://www.fns.usda.gov/cnd/lunch/ aboutlunch/NSLPFactSheet.pdf.

[166] FNS, "Federal Cost of School Food Programs, data as of July 26, 2012," http://www.fns. usda.gov/pd/cncosts.htm. Primary funding sources for USDA food purchases include (as amended): Section 6 of the Richard B. Russell National School Lunch Act; Section 32 of the Act of August 24, 1935; and Section 416 of the Agricultural Act of 1949.

[167] Ibid. Consisted of $10.105 billion in cash assistance and $1.036 billion in donated food assistance.

[168] FNS, "School Food Purchase Study-III Final Report," March 2012, http://www.fns.usda.gov/ Ora/menu/Published/ CNP/FILES/SFSPIII_Final.pdf.

[169] For more information, see CRS Report R41354, Child Nutrition and WIC Reauthorization: P.L. 111-296. Other resources related to the updated guidelines and compliance requirements are at FNS' website: http://www.fns.usda.gov/ cnd/Governance/Legislation /nutritionstandards.htm.

[170] See, e.g., "Rabobank Says Value-Added Fresh Vegetables Positioned for Growth," March 10, 2012, https://www. rabobankamerica.com/content/documents/news/ 2012/value_ added_fresh_vegetables_positioned_for_growth.pdf.

[171] P.L. 110-246, § 4303, amending the NSLA; 42 U.S.C § 1769 (CFDA# 10.582). USDA websites: http://www.fns.usda.gov/cnd/ffvp/ and http://www.fns.usda.gov/cnd/ ffvp/ handbook.pdf. The 2002 farm law included authority to use funding for dried fruits and vegetables, while the expanded and extended program does not include these products.

[172] The annual grant include a minimum grant amount (1% of the funds made available), with an additional allocation to each state based on its population share. Priority is given to schools

where more than 50% of the students are eligible for free or reduced price meals. See also FNS, http://www.fns.usda.gov/cnd/ffvp/.

[173] WIC food packages and nutrition education are primary means by which the program affects the dietary quality and habits of participants. For more information, see USDA, "WIC Food Packages," http://www.fns.usda.gov/wic/ benefitsandservices/foodpkg.htm.

[174] For an overview of SNAP, see CRS Report R42505, Supplemental Nutrition Assistance Program (SNAP): A Primer on Eligibility and Benefits.

[175] FNS, Availability of Fresh Produce in Nutrition Assistance Programs, CN-02-FV, May 2002, http://www.fns.usda.gov/Ora/menu/Published/CNP/FILES/FreshProduce.pdf.

[176] See FNS, "Increasing Fruit and Vegetable Consumption through the USDA Nutrition Assistance Programs," March 2008, http://www.fns.usda.gov/ora/menu/ Published/ NutritionEducation/Files/fruit_veggie_report.pdf.

[177] For information see USDA, FNS, "SNAP: Learn How You Can Accept SNAP Benefits At Farmers' Markets," http://www.fns.usda.gov/snap/ebt/fm.htm. For other information, see CRS Report R42155, The Role of Local Food Systems in U.S. Farm Policy.

[178] USDA FNS, "SNAP's Benefit Redemption Division (BRD) Annual Report for Fiscal Year 2010," May 3, 2011, http://www.fns.usda.gov/snap/retailers/pdfs/2010-annual-report.pdf. Based on a total out of 6,132 farmers' markets.

[179] The FNS provides grants to state agencies, such as state health, agriculture and other agencies and Indian Tribal Organizations (ITOs), in nearly all states. A map of participating states is at http://www.fns.usda.gov/wic/SFMNPFMNP-Map.pdf. Participating state agencies must submit a plan describing how the agency intends to implement, operate and administer the program. Grant payments are made by a letter of credit, and state agencies may withdraw funds only as needed.

[180] OFPA was enacted as part of the 1990 farm bill (P.L. 101-624). NOP regulations are at 7 C.F.R. 205.

[181] USDA, 2008 Organic Production Survey, Volume 3, AC-07-SS-2, July 2010.

[182] Specifically, Title X of the 2008 farm bill is titled "Horticulture and Organic Agriculture," and covers fruits, vegetables, and other specialty crops and organic agriculture (even though organic agriculture includes products other than fruits and vegetables, including meat and poultry products, milk and dairy foods, and field crops).

[183] USDA, USDA Organic Resource Guide 2012, Your Guide to Organic and Organic-Related USDA Programs, August 2012, http://www.ams.usda.gov/AMSv1.0/ getfile?dDocName= STELPRDC5100093.

[184] For information, see USDA's website (http://www.ams.usda.gov/AMSv1.0/nop). See also NOP's "Fact Sheet," NOP's "2010-2012 Strategic Plan," and "National Organic Program," http://www.ams.usda.gov/AMSv1.0/organicinfo.

[185] The NOP regulations prohibit the use of genetic engineering, irradiation, and sewage sludge in certified organic production and handling.

[186] USDA, "2013 Explanatory Notes, AMS," http://www.obpa.usda.gov/19ams2013notes.pdf, p. 19-89. For a history of actual funding since the program began in 2002, see presentation by Miles McEvoy (AMS), "National Organic Program," October 25, 2010, http://www.ams.usda.gov/AMSv1.0/getfile?dDocName=STELPRDC5087263.

[187] P.L. 110-246, § 10303.

[188] However, USDA's program that accredits certification agents is fee-based, allowing accredited certification agents to charge fees to producers (cost-reimbursement basis) for organic certification services (7 CFR § 205.640).

[189] Authorized in the 2002 farm bill (107-171, § 10606), as amended. 7 U.S.C. § 6523 (CFDA#10.171).
[190] Authorized under the Federal Crop Insurance Act (FCIA), as amended (Title 7, Part 36, Section 1501-1524). 7 U.S.C §§ 1501-1524 (CFDA# 10.163). See also USDA's website (www.ams.usda.gov/NOPCostSharing).
[191] 76 Federal Register 54999-55000, September 6, 2011, http://www.gpo.gov/fdsys/pkg/FR-2011-09-06/pdf/2011- 22611.pdf and http://www.gpo.gov/fdsys/pkg/FR-2011-09-06/pdf/2011-22613.pdf; and USDA/AMS, NOP Cost-Share Programs 2012 Report to Congress, http://www.ams.usda.gov/AMSv1.0/getfile?dDocName=STELPRDC5097859.
[192] A list of state contacts is available from USDA (http://www.ams.usda.gov/NOPCostShareProgramParticipants) and the National Association of Organic Programs (www.nasda.org/nasop/).
[193] USDA/AMS, NOP Cost-Share Programs 2012 Report to Congress. Table 1.
[194] P.L. 110-246, § 10301 (amended P.L. 107-171, § 10606, which provided a one-time amount of $5 million).
[195] As reported in various AMS reports to Congress, NOCCSP allocations were: $4.9 million (FY2003-FY2008); $5.9 million (FY2009); $4.8 million (FY2010); $5.0 million (FY2011); and $5.7 million (FY2012). Allocations under NOCCSP ranged from $5,000 to $1,050,000 per state/territory.
[196] USDA, "2013 Explanatory Notes," http://www.obpa.usda.gov/19ams2013notes.pdf (p. 19-10). Allocations under AMA ranged from $15,000 to $450,000 per state.
[197] USDA/AMS, NOP Cost-Share Programs 2012 Report to Congress. Table 2 and Table 3.
[198] P.L. 110-246, § 10107, amending P.L. 107-171, § 7407. 7 U.S.C. 5925c.
[199] USDA, "2013 Explanatory Notes," http://www.obpa.usda.gov/19ams2013notes.pdf, p. 19-89.
[200] Authorized in the Agricultural Research, Extension, and Education Reform Act of 1998 (AREERA, P.L. 105-185, § 244), which amended the 1990 farm bill. 7 U.S.C. § 5925b (CFDA# 10.307).
[201] USDA's website (http://www.nifa.usda.gov/fo/organicagricultureresearchandex tensioni nitiative.cfm/) under "Abstracts of Funded Projects" (recipient names, award amount, and project terms).
[202] AREERA § 406, as amended, established the "Integrated Research, Education, and Extension Competitive Grants Program," of which ORG is a part. 7 U.S.C. § 7626 (CFDA# 10.303).
[203] USDA's website (http://www.csrees.usda.gov/fo/organictransitionsprogram.cfm) under "Abstracts of Funded Projects" (recipient names, award amount, and project terms).
[204] USDA, NIFA, "2013 Explanatory Notes," http://www.obpa.usda.gov/17nifa2013notes.pdf p. 17-74.
[205] EQIP was originally authorized in the 1985 farm bill ((P.L. 99-198), as amended; 16 U.S.C. 3830 (CFDA# 10.912). General information about EQIP is at http://www.nrcs.usda.gov/wps/portal/nrcs/main/national/programs/financial/eqip.
[206] NRCS, "EQIP Organic Initiative," http://www.nrcs.usda.gov/wps/portal/ nrcs/detailfull/national/programs/?cid= nrcs143_008224.
[207] NRCS, "Organic Initiative, 2011 Year-End Review," http://www.nrcs.usda.gov/Internet/FSE_DOCUMENTS/ stelprdb1046662.pdf.
[208] CRS communication with NRCS staff, March 2012. Data are for FY2010.
[209] Including Illinois, Indiana, Michigan, Wisconsin, Iowa, Minnesota, and Ohio.
[210] For more information, see CRS Report RL34019, Eliminating the Planting Restrictions on Fruits and Vegetables in the Farm Commodity Programs.

[211] For information on these programs, see CRS Report R40763, Agricultural Conservation: A Guide to Programs.

[212] S. Zahniser, et al., The Potential Impact of Changes in Immigration Policy on U.S. Agriculture and the Market for Hired Farm Labor: A Simulation Analysis, ERR-135, May 2012, USDA/ERS, http://www.ers.usda.gov/amber-waves/2012-june/immigration-policy. aspx.

[213] UFPA, "Labor and Immigration Reform," http://www.unitedfresh.org/newsviews/ immigration.

[214] Named after the authorizing section in the Immigration and Naturalization Act (INA) of June 27, 1952, § 101(a)(15)(H)(ii)(a)), 8 U.S.C. §1101 et seq. (CFDA#17.273). The INA, as amended, is the basis of current immigration law. DOL's website (http://www.foreign laborcert.doleta.gov/h-2a.cfm).

[215] See CRS Report R42434, Immigration of Temporary Lower-Skilled Workers: Current Policy and Related Issues.

[216] DOL's Employment Law Guide (for H-2A visas) is at http://www.dol.gov/asp/programs/ guide/taw.htm.

[217] Authorized by the Workforce Investment Act of 1998 (WIA, P.L. 105-220, § 167), 29 U.S.C. § 1912 (CFDA#17.264). Related programs started under the Economic Opportunity Act of 1964. Regulations are at 20 CFR Part 652 and 660-671. ETA's website (http://www. doleta.gov/msfw/).

[218] 29 U.S.C. § 1801 et seq. (CFDA#17.308 and 17.303).

[219] See CRS Report RL31325, The Federal Migrant Education Program as Amended by the No Child Left Behind Act of 2001.

In: Specialty Crops
Editor: Katelin Fransen

ISBN: 978-1-62618-054-3

Chapter 2

FEDERAL CROP INSURANCE FOR SPECIALTY CROPS: BACKGROUND AND LEGISLATIVE PROPOSALS*

Dennis A. Shields

SUMMARY

The federal crop insurance program provides farmers with risk management tools to address crop yield and/or revenue losses on their farms. Farmers can purchase subsidized policies that pay an indemnity when their production or revenue falls below a guaranteed level.

Historically, the federal crop insurance program primarily has covered traditional farm program crops such as wheat, corn, and soybeans. However, the crop insurance program has expanded in recent decades, and policies are available now for a wide range of commodities, including specialty crops such as fruits and vegetables. In 2009, specialty crop policies covered more than 7 million acres, which was roughly one-half to three-quarters of specialty crop area, depending on how total area is calculated. In total, crop insurance is available for over 80 specialty crops, making the program the primary financial safety net for specialty crop producers.

* This is an edited, reformatted and augmented version of the Congressional Research Service Publication, CRS Report for Congress R42813, dated November 6, 2012.

The federal crop insurance program is permanently authorized by the Federal Crop Insurance Act, as amended (7 U.S.C. 1501 et seq.). The U.S. Department of Agriculture's (USDA's) Risk Management Agency (RMA) operates the Federal Crop Insurance Corporation (FCIC), which is the funding mechanism for the program. Insurance policies are sold and completely serviced through approved private insurance companies. The insurance companies' losses are reinsured by USDA, and their administrative and operating costs are reimbursed by the federal government.

Federal crop insurance policies for specialty crops (and other crops as well) are generally either yield-based or revenue-based. For most yield-based policies, a producer can receive an indemnity if there is a yield loss relative to the farmer's "normal" (historical) yield. Insurable causes of loss include drought, excess precipitation, hail, frost, freeze, fire (if due to natural causes), and insects and disease. Revenue-based policies protect against crop revenue loss resulting from declines in yield, price, or both. Nursery crop producers can be protected against loss of an "asset" due to weather damage (but not price loss). Also relevant for specialty crop producers is whole-farm insurance, which protects against losses in whole farm revenue rather than just for an individual crop. An endorsement is available in some areas for protecting against loss due to quarantine.

While additional policies have been introduced in the last 10 years, producers and some Members of Congress would like to enhance crop insurance for specialty crop producers. In the 112th Congress, several legislative proposals have been introduced. The Local Farms, Food, and Jobs Act of 2011 (H.R. 3286/S. 1773) would enhance whole farm insurance and insurance for organic crops. The Rural Economic Farm and Ranch Sustainability and Hunger (REFRESH) Act of 2011 (S. 1658 and H.R. 3111) would also enhance whole farm insurance. The Specialty Crop Insurance Act of 2011 (S. 1905) would support the development of new policies.

In the 112th Congress, specialty crop provisions in these and other proposals have been included in the Senate-passed farm bill (S. 3240) and the House committee-passed farm bill (H.R. 6083). The bills require USDA to conduct more research on higher coverage levels for whole farm revenue insurance and study insurance for food safety and contamination-related losses. A provision would revise the value of crop insurance for all organic crops to reflect prices of organic (not conventional) crops. For crops not insurable with crop insurance, additional coverage would become available under the existing Noninsured Crop Assistance Program (NAP). Finally, a provision in S. 3240 would subsidize the purchase of existing private-sector index-based weather insurance, which protects against specific weather events and not actual loss.

Introduction

The federal crop insurance program provides producers with risk management tools to address crop yield and/or revenue losses on their farms. Farmers can purchase subsidized policies that pay an indemnity when their production or revenue falls below a guaranteed level.

The crop insurance program has expanded in recent decades, and policies are available now for a wide range of commodities, including crops covered by federal farm programs (e.g., wheat and corn) as well as specialty crops such as fruits and vegetables. While additional policies have been introduced in the last 10 years, producers and some Members of Congress would like to enhance crop insurance for specialty crop producers. Several legislative proposals have been introduced in the 112th Congress, including provisions in the Senate-passed farm bill (S. 3240) and the House Agriculture Committee-reported farm bill (H.R. 6083).

This report describes the federal crop insurance program for specialty and organic crops.[1] In 2009, specialty crop policies covered more than 7 million acres, which was roughly one-half to three-quarters of specialty crop area, depending on how total area is calculated. In total, crop insurance is available for over 80 specialty crops.[2]

Crop Insurance Program Authority and Operation

The federal crop insurance program is permanently authorized by the Federal Crop Insurance Act, as amended (7 U.S.C. 1501 et seq.). It is periodically modified, most recently in the 2008 farm bill (P.L. 110-246). Congress chose to revise the legislation in the 2008 farm bill to achieve budget savings and to supplement crop insurance with a permanent disaster payment program.

The U.S. Department of Agriculture's (USDA's) Risk Management Agency (RMA) operates the Federal Crop Insurance Corporation (FCIC), which is the funding mechanism for the program. Management is vested in a board of directors, subject to the general supervision of the Secretary of Agriculture. Insurance policies are sold and completely serviced through approved private insurance companies. Independent insurance agents are paid sales commissions by the companies. The insurance companies' losses are reinsured by USDA, and their administrative and operating costs are reimbursed by the federal government.

In purchasing a policy, a producer growing an insurable crop selects a level of coverage and pays a portion of the premium, which increases as the level of coverage rises. The remainder of the premium is covered by the federal government (about 60% of total premium, on average, is paid by the government).[3] In the case of catastrophic coverage, the government pays the full premium. (For more information, see the section below on "Crop Insurance Premiums and Subsidies.") In the absence of subsidies, farmer participation in the crop insurance program would be substantially lower. A major benefit for producers is the timely payment for crop losses (about 30 days after the farmer signs the claim form).

The availability of crop insurance for a particular crop in a particular region is an administrative decision made by USDA. The decision is made on a crop-by-crop and county-by-county basis, based on farmer demand for coverage and the level of risk associated with the crop in the region, among other factors. In areas where a policy is not available, farmers may request that RMA expand the program to their county. The process usually starts with a pilot program in order for RMA to gain experience and test the program components before it becomes more widely available. Alternatively, a policy can be reviewed and later discontinued if it fails to perform at an acceptable level (e.g., low participation or high losses). RMA also regularly responds to requests from commodity organizations or industry representatives for enhancements to existing coverage, such as adding revenue coverage.

CROP INSURANCE HISTORY AND COVERAGE

Congress first authorized federal crop insurance as an experiment to address the effects of the Great Depression and crop losses seen in the Dust Bowl. In 1938, the FCIC was created to carry out the program, which focused on major crops like wheat in major producing regions. During the same era, farm programs were established to support crop prices and boost farm income for producers of so-called program crops, including wheat, corn, and cotton. The availability of federal crop insurance remained limited until passage of the Federal Crop Insurance Act of 1980 (P.L. 96-365), which expanded crop insurance to many more crops and regions of the country. Congress again enhanced the crop insurance program, including greater subsidy levels, in 1994 and in 2000 in order to encourage greater participation. The changes also expanded the role of the private sector in developing new products that would help farmers manage their risks. Today, many banks, when making operating loans, require that farmers purchase crop insurance.[4]

By 1999, insurance policies covered 52 specialty crops with planned testing on another 9, according to a government report.[5] These 61 crops reportedly accounted for a majority of value for specialty crops, but coverage on about 300 additional crops remained unavailable. The report concluded that progress on expansion was slow because USDA follows a deliberate and involved process (taking up to five years) to assess risk and set premiums to ensure that programs are actuarially sound, meaning that total premiums including subsidies cover indemnities. (USDA is required by law to achieve actuarial soundness.) Expansion continued over the subsequent decade. By 2011, insurance was available for more than 80 specialty crops in counties considered to be major growing areas. Crops include avocados, blueberries, grapes, citrus, onions, pumpkins, and tomatoes, among others.[6]

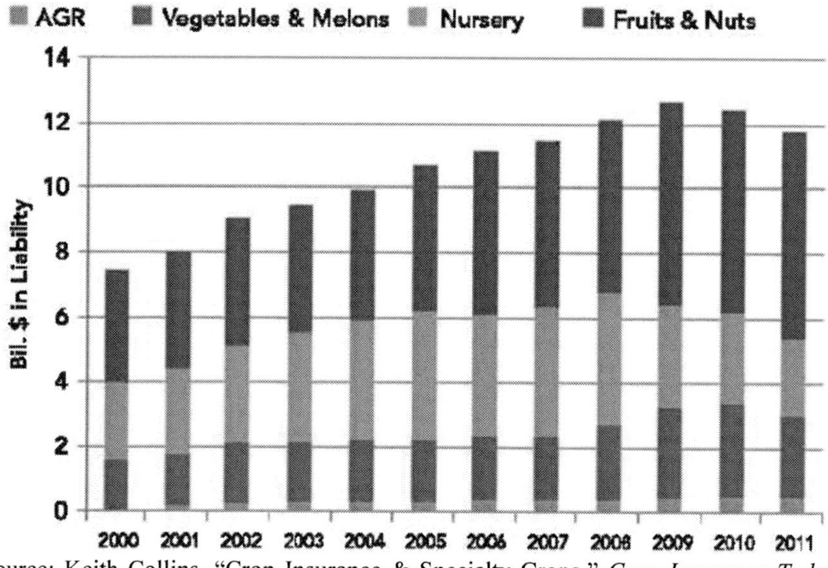

Source: Keith Collins, "Crop Insurance & Specialty Crops," *Crop Insurance Today*, August 2012, pp. 35-40, using data from USDA/RMA.

Figure 1. Insured Liability of Specialty Crops.

Another measure of program growth is the total liability, which is the estimated value of the insured portion of the crop (Figure 1). Insured liability of the specialty crops rose from less than $8 billion in 2000 to nearly $13 billion in 2009, reflecting increased production and participation. Liability declined some in recent years, primarily because of reduced nursery insurance following the severe recession that adversely affected the housing market and demand for nursery products.[7]

When crop insurance is not available, USDA's noninsured crop disaster assistance program (NAP) provides catastrophic coverage (i.e., indemnifying for yield losses greater than 50%) if purchased by the producer. NAP applicants pay an administrative fee (currently $250 per crop); no premiums are required. NAP is administered by the Farm Service Agency, separately from crop insurance.[8]

Development of New Insurance Policies

Producer interest in policy availability often starts at the local level and is channeled through RMA's regional offices. In general, for a viable policy, a crop must have established cultivars, defined farming practices, developed markets, and known perils. Significant producer interest (demand for the policy) also is critical for success.

The Federal Crop Insurance Act provides two methods for developing new crop insurance programs: (1) internal products developed under contract, and (2) external products submitted through procedures in Section 508(h) of the Federal Crop Insurance Act. Once developed, products undergo a rigorous approval process, which can take up to a year to complete. An independent external panel of experts reviews potential policy and actuarial weakness and suggests improvements. The revised product is submitted to the FCIC board of directors for approval. Once approved, the product is typically implemented as a pilot program in a limited area to test for effectiveness while limiting financial exposure. Pilot programs typically operate for four years but may be extended for additional testing if needed. Eventually the board either converts the pilot to a regular program or terminates it.

Sometimes product development is suspended. For example, in recent years, RMA began investigating a product called "Named Peril Weather Insurance," which would allow producers to insure against specific weather events. However, the agency suspended its work until it can be shown that private industry, which offers weather-based products, is not meeting the needs of specialty crop producers. Under law, RMA is not allowed to conduct pilot programs if insurance against the risk is generally available from private companies without public involvement.

TYPES OF POLICIES FOR SPECIALTY CROPS

Several types of federal crop insurance policies are available for specialty crops.

- For most yield-based policies, a producer can receive an indemnity if there is a yield loss relative to the farmer's "normal" (historical) yield. Insurable causes of loss include drought, excess precipitation, hail, frost, freeze, fire (if due to natural causes), and insects and disease.[9]
- Revenue-based policies protect against crop revenue loss resulting from declines in yield, price, or both.
- Whole-farm insurance protects against losses in whole farm revenue rather than just for an individual crop.
- An endorsement is available in some areas for protecting against loss due to quarantine established to control a specific pest that requires destruction of insured crops.

Yield-Based Insurance

Actual production history (APH) and dollar plans are the two major types of yield-based policies for specialty crops. When purchasing an APH policy, a producer is assigned (1) a "normal" crop yield based on the producer's actual production history, and (2) a price for the commodity based on estimated market conditions. The producer can then select a percentage of his normal yield to be insured and a percentage of the price he wishes to receive when crop losses exceed the selected loss threshold.[10]

The most basic level of insurance is called catastrophic (CAT) coverage. The premium is completely subsidized by the federal government. The farmer pays an administrative fee ($300 per crop per county), and in return can receive a payment on losses in excess of 50% of normal yield, equal to 55% of the estimated market price of the crop (called 50/55 coverage). Coverage levels that are higher than CAT are called "buy-up" or "additional" coverage. For an additional premium paid by the producer, and partially subsidized by the government, a producer can "buy up" the 50/55 catastrophic coverage to any equivalent level of coverage between 50/100 and 75/100 (i.e., up to 75% of "normal" crop yield and receive 100% of the estimated market price). In limited areas, production can be insured up to the 85/100 level of coverage. APH policies are popular for a wide variety of specialty crops, including apples, stone fruits, and some citrus fruits.[11] The dollar plan provides protection against declining value due to damage that causes a yield shortfall. The policy guarantees a dollar amount of coverage and not a level of production, with the amount of insurance based on the cost of growing a crop in a specific area. A loss occurs when the annual crop value is less than the amount of insurance. The maximum dollar amount of insurance is stated on the actuarial document. The insured may select a percentage of the maximum dollar amount equal to CAT (catastrophic level of coverage), or additional coverage levels. Dollar plan insurance is popular for nursery crops and citrus. In general, crops are insurable only in certain states and certain counties within those states.

Specialty Crop Insurance Examples: Actual Production History (APH) and Dollar Plan

Actual Production History (APH) Policy for Citrus Fruit

A claim can be filed whenever production falls short of the guarantee selected by the insured. The indemnity is determined by multiplying the production shortfall by the pre-selected price. The following example is based on an actual production history (APH) yield of 600 cartons per acre, 50% coverage level, and a selected price of $5.55 per carton.

600	Cartons per acre average yield (APH yield)
x 0.50	Selected coverage level percentage
300	Cartons per acre guarantee
- 200	Cartons per acre actually produced
100	Cartons per acre loss
x $5.55	Price Election
$555	**Gross indemnity per acre**

Source: USDA Fact sheet, "Citrus–Actual Production History," June 2012, http://www.rma.usda.gov/fields/ca_rso/2012/azcitrus.pdf.

Dollar Plan for Nursery

The dollar plan provides protection against declining value due to damage that causes a yield shortfall.

$100,000	Plant inventory value
x 0.65	Selected coverage level percentage
$65,000	Unit amount of insurance
In the event of a loss:	
$100,000	Field market value before loss
-$50,000	Field market value after loss
$50,000	Value of loss
-$35,000	Deductible (1.0 - coverage level) X inventory = (1 - .65) X $100,000 = $35,000
$15,000	**Indemnity**

Source: USDA Fact Sheet, "Nursery Commodity Insurance," April 2009, http:// www.rma.usda.gov/pubs/rme/nursery.pdf.

Nursery crop insurance is available in all states, but the eligible business enterprise must derive at least 50% of its gross income from the wholesale marketing of plants. The nursery program is an asset-based form of insurance that insures against declines in asset values rather than insuring a percentage of average historical yield or revenue.[12] More than 20,000 plants are listed as insurable, and coverage is available for either field-grown or container products.[13]

The American Nursery and Landscape Association has said that the nursery industry considers the federal crop insurance program to be a valuable component of risk management practices. According to the association, incremental program enhancements have been made in recent years, including the use of a grower's wholesale price list for the basis of coverage. However, they also contend that the program is challenged by the extreme diversity and unique situations found in the industry because production practices are less uniform than for major field crops.

Revenue-Based Insurance

Revenue insurance is widely available for farm program crops (e.g., wheat and corn). For specialty crops, designing revenue products is challenging because the commodities often lack a centralized price discovery mechanism such as a futures exchange for developing price projections prior to planting and actual harvest-time prices needed to determine the insurance guarantee and potential grower indemnity.

Specialty Crop Insurance Example: Actual Revenue History (ARH) for California Navel Oranges

ARH protects against losses from low yields, low prices, low quality, or any combination of these events. Coverage is based on the farmer's own net revenue history and the farmer's choice of coverage level and payment factor (which reduces the premium cost and potential indemnity). The crop's revenue is determined after harvest at the point of first delivery.

$3,200	Per acre approved revenue
x 0.75	Selected coverage level percentage
x 1.0	Selected payment factor (ranges from 0.67 to 1.0 to reduce premium cost to producer
$2,400	Per acre amount of insurance
400	Cartons
x 5.00	Per carton net price
$2,000	Per acre revenue to count
$2,400	Per acre amount of insurance
-$2,000	Per acre revenue to count (actual revenue)
$400	Difference
x 1.0	Payment factor
$400	**Indemnity per acre**

Source: USDA Fact Sheet, "Actual Revenue History-Navels," March 2012, http://www.rma.usda.gov/fields/ca_rso/2012/caarhnavels.pdf.

To address these data challenges, a plan called actual revenue history (ARH) has been implemented on a pilot basis for a few crops such as navel oranges and cherries. Rather than insuring historical yields, the policy insures historical revenues using historical prices.

This approach assumes that historical prices provide a reasonable estimate of expected prices (the assumption is less tenable with storable crops where stock carryover from the previous year can affect current market-year prices).

In its November 2010 report to Congress, the Risk Management Agency indicated that developing revenue coverage for additional specialty crops remains a high priority.

Whole Farm Insurance

Adjusted Gross Revenue (AGR) and AGR-Lite policies insure revenue of the entire farm rather than an individual crop. AGR first appeared in 1999 to protect against production or market losses. Compared with AGR, AGR-Lite has higher levels of coverage available for producers who have multiple commodities. The plan uses a producer's five-year historical farm average revenue as reported on the Internal Revenue Service (IRS) tax return form (Schedule F or equivalent forms) and an annual farm report as a base to provide a level of guaranteed revenue for the insurance period (a one-year period corresponding with the producer's IRS tax period). Coverage levels range from 65% to 80% of historical revenue.[14]

SPECIALTY CROP INSURANCE: WHOLE FARM INSURANCE EXAMPLE

Adjusted Gross Revenue (AGR) and Adjusted Gross Revenue-Lite (AGR-Lite) are whole-farm, revenue-protection plans of insurance. Compared with AGR, AGR-Lite has higher coverages available for producers who have multiple commodities.

$100,000	Approved adjusted gross revenue
x 0.80	Selected coverage level percentage
$80,000	Amount of insurance

In the event of a loss:

$80,000	Amount of insurance
-$70,000	Revenue to count (actual revenue)
$10,000	Revenue to count
x 0.75	Selected payment rate
$7,500	**Indemnity**

Source: USDA Fact Sheet, "Adjusted Gross Revenue-Lite," November 2010, http://www.rma.usda.gov/pubs/rme/agr- lite.pdf.

In general, the AGR products are designed to protect producers with specialty crops and/or commodities not covered by individual policies. Historically, though, whole-farm insurance has seen limited participation. With individual crop insurance policies already providing significant protection for many producers, combined sales of AGR and AGR-Lite were less than 1,000 policies in 2011, a small fraction of the 2 million crop insurance policies sold. Also, observers say the AGR products are complicated in terms of compiling the information needed to consider purchasing the insurance and making the application. Others also have noted that for such a policy to be widely adopted, coverage levels need to be substantially higher than individual crop insurance policies (i.e., higher than the current 80% level) in order to provide an amount of risk protection equivalent to that afforded by individual crop policies. A delay in indemnity payment also has been cited by producers as a drawback to those policies.

Quarantine Endorsement

A quarantine endorsement is available for California avocado and citrus, whereby producers may purchase additional coverage to insure against losses associated with a quarantine established by USDA or the California Department of Food and Agriculture. The quarantine would be established to control a specific pest that requires the destruction of insured crops or plants on when the insured crop is growing, or does not permit the insured crop to be harvested, sold, or transported or otherwise restricts its movement from the farm.[15]

Insured Acreage and Participation Rates

In its report to Congress on specialty crop insurance, FCIC identified more than 7 million acres of specialty crops that were enrolled in federal crop insurance in 2009, or an average participation rate of 75% for the 34 crops shown in Table 1. (If all specialty crop acreage is included in total crop area (i.e., not just insurable crops), the insured share of total crop acreage is equal to 53%.)[16] This compares with a participation rate of 83% for major program crops (i.e., those under farm commodity price and income support programs)[17.] Insured acreage as a share of crop acreage is relatively high in major specialty crop states, including California (71% of total crop area), Florida (91%), and Washington (68%).

Acreage participation for pulse crops (e.g., dry peas, dry beans) is high in the major growing states in the Northern Plains, including Minnesota (84%), Montana (83%), North Dakota (96%), and South Dakota (79%).

Table 1. Specialty Crop Insurance: Participation Rate and Liability, 2009

Selected crop	Insured acres	Total acres	Participation rate	Liability ($)
Total	**7,252,820**	**9,661,557**	**75%**	**8,954,543,672**
Raisins	249,160	249,160	100%	136,735,740
Prunes	61,752	64,000	96%	76,737,020
Citrus fruit	786,286	857,041	92%	777,554,804
Table grapes	84,691	97,000	87%	197,015,733
Dry beans	1,276,208	1,501,511	85%	429,497,134
Dry peas	999,132	1,171,000	85%	204,663,983
Cranberries	31,616	38,500	82%	147,889,706
Tomatoes	326,893	409,250	80%	677,197,421
Macadamia nuts	11,837	15,000	79%	22,686,079
Potatoes	834,062	1,065,350	78%	1,061,826,089
Citrus trees	21,144	27,300	77%	1,358,916,773
Green peas	172,786	224,380	77%	65,429,637
Plums	21,795	29,500	74%	29,838,870
Nectarines	23,009	32,300	71%	25,983,541
Almonds	478,380	680,000	70%	935,087,199
Peaches	84,354	122,609	69%	125,084,666
Apples	236,756	348,200	68%	694,494,268
Grapes	567,492	838,310	68%	796,391,315
Figs	6,002	9,400	64%	3,674,407
Pears	35,104	56,000	63%	63,255,712
Cherries	49,384	81,910	60%	279,906,372
Apricots	7,141	12,380	58%	11,464,376
Pecans	156,973	279,450	56%	126,722,514
Sweet corn	319,746	577,795	55%	152,196,648
Blueberries	44,055	81,536	54%	91,719,749
Avocados	37,781	72,500	52%	65,087,114
Onions	80,040	160,971	50%	139,283,810
Processing beans	102,542	209,324	49%	44,595,759
Walnuts	110,071	223,000	49%	159,823,397
Peppers	9,204	19,000	48%	31,911,660
Pumpkins	6,188	13,100	47%	3,489,526
Tropical fruit	3,784	8,780	43%	975,117
Chile peppers	3,771	15,800	24%	2,009,452
Cabbage	13,681	70,200	19%	15,398,081

Source: Columns 1-3 from “Report to Congress: Specialty Crop Report,” USDA, Federal Crop Insurance Corporation, November 2010, http://www.rma.usda.gov/pubs/2010/specialtycrop.pdf. Column 4 from USDA, Risk Management Agency summary of business reports and data.

Note: Not included in total is liability for nursery products at $3.2 billion (acreage not applicable).

Other states with specialty crop production (and their participation rates) include Michigan (73%), New York (70%), and Oregon (52%). The USDA report contains detailed acreage data by crop and state, along with maps showing crop insurance participation.

CROP INSURANCE PREMIUMS AND SUBSIDIES

For crop insurance policies, the premium increases as the level of insurable yield and price coverage rise. For "catastrophic" coverage (i.e., yield loss greater than 50%), the premium is 100% subsidized by the federal government. As stated earlier, the farmer pays an administrative fee for CAT coverage ($300 per crop per county), and in return can receive an indemnity on losses in excess of 50% of normal yield, equal to 55% of the estimated market price of the crop (called 50/55 coverage). For protection from greater financial losses, producers can select "buy- up" coverage. The subsidy for these policies ranges from 67% to 38%.[18]

Unlike other forms of insurance, the premium for crop insurance reflects only costs associated with policy risk and does not include company costs for selling and servicing the policies. These are covered by administrative and operating (A&O) reimbursements paid by the government to insurance companies.

The lower level of acreage participation for specialty crops relative to program crops (described in the previous section) is reflected in aggregate and relative levels of producer subsidies for crop insurance premiums. For an aggregate comparison, producer subsidies for federal crop insurance premiums in 2011 totaled $438 million for specialty crops and $6.9 billion for program crops (i.e., those covered by farm commodity programs). The combined total was $7.3 billion.[19]

A relative comparison can be made by contrasting the share of crop insurance subsidies of each crop category with its share of total crop value. The specialty crop share of premium subsidies was 6% while its share of total crop value was 31%. In contrast, the premium subsidy share for program crops was 94% of the total while its share of total crop value was 69%.[20] Another way to view the same data is to calculate a ratio of premium subsidies to crop value for each of the two categories. The ratio is 0.7% for specialty crops, and 5.1% for program crops. These figures illustrate that, relative to total crop value, specialty crops in general are subsidized at lower levels than program crops.

Factors affecting the difference between program and specialty crops in subsidy amounts include acreage participation and crop insurance availability. (The premium subsidy rates are the same for all crops.) Another important factor is the level of coverage purchased, which is significantly lower for specialty crops, particularly fruits and nuts. In 2011, for example, acreage with coverage above the catastrophic level (i.e., 50% deductible) was 59% for specialty crops compared with 89% for program crops. Higher coverage levels purchased by farmers correspond with higher premiums (and associated premium subsidies).

INSURANCE FOR ORGANIC CROPS

USDA reports that over 80 crop varieties were insured either as certified or transitional organic under crop insurance in recent years, with annual area rising from 210,393 acres in 2004 to 565,195 acres in 2009. The department estimates organic crop insurance participation at nearly 35%.[21]

Prior to 2000, when the Agricultural Risk Protection Act (ARPA) was enacted, crop insurance for organic production was available only by written agreement.[22] Under a written agreement, a producer could receive an individual determination of coverage for specific crops in specific counties. Each situation was evaluated separately based on a farmer's organic cropping practices. ARPA allowed organic production to be considered under the terms and conditions of all crop policies available at that time. By 2004, RMA modified the basic provisions such that organic producers were eligible for standard crop insurance policies and were not required to obtain a written agreement.[23]

For organic crops, federal crop insurance is available for (1) certified organic acreage, (2) transitional acreage (acreage on which organic farming practices are being followed but does not yet qualify to be designated as certified organic acreage), and (3) buffer zone acreage (used to minimize the possibility of unintended contact by prohibited substances).[24]

To insure certified organic acreage, a producer must have a current organic plan and written certification from a certifying agent. For transitional acreage, a producer needs written documentation from a certifying agent showing that an organic plan is in effect (acreage without documentation can be insured under conventional farming practices). Separate yield history (4-10 years of data) is maintained for conventional, transitional, and certified organic acreage. Importantly, indemnities are made only where planned controls or practices were not effective.

In general, FCIC charges an additional 5% surcharge on total premium for insuring organic crops to cover additional risk associated with organic farming practices. However, premium adjustments are reviewed annually, and can be eliminated based on the accumulated insurance experience.

For the 2011 crop year, no additional charge was required for the following organic crops: figs, Florida citrus fruit, Florida fruit trees, macadamia trees, nursery, pears, peppers, prunes, Texas citrus trees, and Texas citrus fruit.

A number of plans of insurance do not contain an additional 5% surcharge for organic production because they do not distinguish between organic and conventional production practices. These include AGR, AGR-Lite, Group Risk Plan (which uses county loss experience as the basis for determining loss), and Group Risk Income Protection (which makes an indemnity when average county revenue is below the guarantee level).

Separately from premium charges, for most organic crops, the price elections and price data used for establishing guarantee levels and indemnities are the same as for crops grown under conventional practices. Organic producers have criticized such pricing for not representing the higher value of their commodities. In response, in 2011, USDA began using separate organic price elections and price data for organically produced cotton, corn, soybeans, and processing tomatoes.

INSURANCE ISSUES FOR SPECIALTY CROPS

While adoption of federal insurance policies for specialty crops has increased substantially in recent decades, USDA and the industry continue to face a number of issues when developing and making available new insurance policies for specialty crops currently not covered.[25]

These issues are not new and are generally difficult to address, as they stem from the basic structure of the industry. (Issues with whole farm insurance and organic insurance are covered in previous sections.)

First, the fact that specialty crop acreage for individual crops is small affects the potential marketability of a policy. A small market reduces sales incentive for companies selling insurance while contributing to higher per-unit costs for developing the product, training staff, modifying computer programs, and other activities.

Small acreage also results in low market volume or the establishment of production contracts between producers and buyers. As a result, in the absence of a well-developed cash market, such “thin market” conditions make it difficult to observe and forecast market prices, which affects RMA’s ability to set the appropriate level of price guarantees.

Setting price guarantees correctly is critical for encouraging participation, actuarial soundness of the program, and overall market dynamics for the crop. If the insurance is priced (rated) too high, producers who tend to have few losses might decide against purchasing insurance, leaving only high-risk farmers in the pool (called “adverse selection”).[26] If the insurance is priced too low, premiums may not cover expected indemnities, potentially inflating the federal cost of the program.

An additional concern is that artificially low premiums might encourage additional crop production, resulting in weak market prices and negative financial returns for producers.

Another challenge for specialty crops is the multitude of crop varieties and production practices. Compared with field crops, specialty crops tend to have a wider variety of farming practices that depend on the crop variety, adding complexity to the policy and its development cost. For example, a vegetable crop may need to be grown on raised beds or use plastic, or require specific crop rotations.

Understanding how these factors affect potential yields is required for determining what practices can be insured and for developing and establishing underwriting standards. Variation across crops and by variety within crop types also complicates the loss adjustment process (i.e., assessing the effect of weather on crop production).

LEGISLATIVE PROPOSALS

Several bills introduced in the 112th Congress are designed to enhance crop insurance for specialty crops and address some of the insurance issues facing producers. Given the inherent difficulties (described above) associated with providing crop insurance for specialty crops, they are designed primarily to enhance the product development process in order to cover more commodities or expand and improve whole farm insurance and organic insurance for producers.

Specific proposals include the following.

- **The Local Farms, Food, and Jobs Act of 2011 (H.R. 3286/S. 1773)** was introduced by Representative Pingree and Senator Brown. The bill would require the Federal Crop Insurance Corporation to offer a whole-farm revenue risk plan in all states and counties. Producers would qualify for an indemnity if actual gross farm revenue is below 85% of the average gross farm revenue of the producer (currently limited to 80%). Producers of any type of agricultural commodity would be eligible. In addition, coverage is to include the value of any packing, packaging, labeling, washing or other on-farm activities needed to facilitate sale of the commodity. In addition, like S. 1905 below, the bill would remove the prohibition of FCIC conducting its own insurance product research.
  The bill also contains a section on organic crops. Premium surcharges would be eliminated on insurance policies for organic crops, and FCIC would be required to offer insurance at actual price levels received by growers for all organic crops produced in compliance with standards issued by USDA. Finally, FCIC would be required to submit an annual report to Congress summarizing the varieties of organic crops insured, progress on implementing the required price changes, development of new insurance products relevant to organic producers, and any recommended changes to improve insurance coverage for organic producers.
- **The Specialty Crop Insurance Act of 2011 (S. 1905)** by Senator Gillibrand would waive the 50% cap imposed on the proportion that FCIC can pay in advance to applicants for research and development of new insurance plans. The waiver is meant to support the development of specialty crop plans of insurance for coverage in regions or crops that are considered underserved by the FCIC and when the applicant does not have sufficient financial resources to fund the development of the concept. In addition, the bill would allow FCIC to conduct its own research and development on existing or new policies.
- **The Rural Economic Farm and Ranch Sustainability and Hunger (REFRESH) Act of 2011 (S. 1658 and H.R. 3111)** by Senator Lugar and Representative Stutzman proposes comprehensive changes to current U.S. farm policy. Among the provisions is an expansion of whole-farm insurance. FCIC is instructed to extend whole farm

insurance (as part of a pilot program) to all counties, subject to a rating as determined by FCIC. Currently, USDA offers whole farm revenue insurance in selected states. In addition, the bill would remove the prohibition of FCIC conducting its own insurance product research.

As part of the ongoing debate over the omnibus farm bill, several of these insurance provisions for specialty crops have been included in the Senate-passed farm bill (S. 3240) and the House committee-reported farm bill (H.R. 6083). Selected provisions in both bills with potential benefit for specialty crop producers would

- require USDA to (1) conduct more research on whole farm revenue insurance with higher coverage levels than currently available, and (2) conduct a study on insuring specialty crop producers for food safety and contamination-related loss;
- enhance the Noninsured Crop Assistance Program (NAP), which offers catastrophic coverage for crops not insurable under the crop insurance program, by making available additional coverage at 50% to 65% of established yield and 100% of average price;
- allow producers to purchase a second crop insurance policy called "Supplemental Coverage Option" or SCO, which is designed to cover part of the deductible under the producer's underlying policy; SCO is intended for crops, including specialty crops, that have sufficient data needed for policy development and implementation;
- revise the value of crop insurance for all organic crops to reflect prices of organic (not conventional) crops (specific provision in S. 3240); and
- subsidize the purchase of existing private-sector index-based weather insurance, which insures against specific weather events and not actual loss (S. 3240 only).

These and other provisions in the House committee-reported and Senate-passed farm bills that would affect crop insurance (and disaster assistance) for specialty crops are summarized in Table 2.

Table 2. Crop Insurance and Disaster-Related Provisions for Specialty Crops in a 2012 Farm Bill

Current Law/Policy	Senate-Passed Farm Bill (S. 3240)	House Agric. Comm. Bill (H.R. 6083)
Permanently authorized by the Federal Crop Insurance Act, the federal crop insurance program makes available subsidized crop insurance to producers who purchase a policy to protect against individual farm losses in yield, crop revenue, or whole farm revenue. ***[7 U.S.C. 1501 et seq.]***	Retains current program and makes available to crop producers (including specialty crop producers if data are available) an additional policy called Supplemental Coverage Option (SCO) to cover part of the deductible under the producer's underlying policy. SCO is an area-wide (e.g., county) yield or revenue loss policy, whereby an indemnity is paid on area losses greater than 10% of normal level and not more than the deductible level (e.g., 25%) selected by the producer in the underlying individual policy. Premium subsidized at 70%. Coverage to begin no later than the 2013 crop year. ***[Sec. 11001]***	SCO provision is similar to the Senate bill. ***[Sec. 11003]***
Requires FCIC to improve coverage for organic crops by signing contracts to develop improved policies as well as a procedure to offer higher price elections. Requires annual report to Congress. ***[U.S.C. 1522(c)(10)]***	Deletes provision and requires FCIC to offer price elections for all organic crops by 2015 that reflect prices of organic (not conventional) crops. FCIC must submit an annual report to Congress on crop insurance for organic crops. ***[Sec. 11021]***	Extends 2008 farm bill provision (without specific new provisions) to improve organic crop insurance. ***[Sec. 11021]***
FCIC shall not conduct any pilot program that provides insurance protection against a risk if a policy is generally available from private companies. ***[7 U.S.C. 1523(a)]***	FCIC may conduct a pilot program to provide financial assistance for producers of underserved crops and livestock (including specialty crops) to purchase an index-based weather insurance product from a qualified private insurance company. The subsidy shall not exceed 60% of the estimated premium amount. Unlike FCIC policies, the private insurance companies would maintain exclusive rights to rate and manage the policies. Provides mandatory funds of $10 million per year for FY2013 through FY2017. ***[Sec. 11024]***	No comparable provision.

Current Law/Policy	Senate-Passed Farm Bill (S. 3240)	House Agric. Comm. Bill (H.R. 6083)
Under Sections 522 and 523 of the Federal Crop Insurance Act, FCIC may enter into contracts to carry out research and development for new crop insurance policies (but may not conduct research itself). FCIC shall establish as one of the highest research priorities the development of a pasture, range, and forage program. It shall provide a payment to an applicant for research and development costs. FCIC may approve up to 50% of the projected total research and development costs to be paid in advance to an applicant. ***[7 U.S.C. 1522(c)]***	Allows FCIC to conduct research and development activities to maintain or improve existing policies or develop new policies. Highest research priorities become policies that increase participation by producers of underserved agricultural commodities, including sweet sorghum, sorghum for biomass, specialty crops, sugarcane, and dedicated energy crops ***[Sec. 11022]***	Same as current law (i.e., does not provide authority for FCIC to conduct research itself) but revises high priority research topics as specified in Senate bill. ***[Sec. 11020]***
	FCIC shall review any policy developed under Section 522(c)or any pilot program developed under Section 523 and submit the policy or program to the Board if it finds that the policy or program will likely result in a viable and marketable policy and would provide coverage in a significantly improved form. ***[Sec. 11007]***	Identical to the Senate bill. ***[Sec. 11010]***
	For cost reimbursement, the 50% limitation may be waived and, upon request of the submitter, an additional 25% advance payment may be made. ***[Sec. 11015]***	No comparable provision.
	FCIC is required to contract for studies on the feasibility of insuring specialty crop producers for food safety and contamination-related losses ***[Sec. 11017]***	Identical to the Senate bill. ***[Sec. 11021]***
Adjusted Gross Revenue (AGR) and AGR-Lite policies insure revenue of the entire farm rather than an individual crop. Both use a producer's five-year historical farm average revenue as reported on the Internal Revenue Service (IRS) tax return form (Schedule F or equivalent forms). Coverage levels range from 65% to 80% of historical revenue. ***[7 U.S.C. 1523]***	FCIC is to conduct activities or enter into contracts to develop a whole farm risk management insurance plan (with liability up to $1.5 million) that pays an indemnity if gross farm revenue is below 85% (compared with 80% currently). Coverage may include value of packing, packaging or other on-farm activities. FCIC may provide diversification-based discounts for producers with diversified operations. ***[Sec. 11016]***	Identical to the Senate bill, except maximum liability is $1.0 million. ***[Sec. 11021]***

Table 2. (Continued)

Current Law/Policy	Senate-Passed Farm Bill (S. 3240)	House Agric. Comm. Bill (H.R. 6083)
The Standard Reinsurance Agreement (SRA) between FCIC and private companies defines expense reimbursements and risk-sharing by the government, including the terms under which the government provides subsidies and reinsurance (i.e., insurance for insurance companies) on eligible crop insurance contracts sold or reinsured by insurance companies. FCIC may renegotiate the SRA once every 5 years. ***[7 U.S.C. 1508(k)]***	Any savings generated from a renegotiated SRA must be used for programs administered by the Risk Management Agency. ***[Sec. 11010]***	Same as Senate bill. [Sec. 11012] Also directs FCIC to make an additional annual expense reimbursement of $41 million (for reinsurance years 2011 through 2015) to insurance companies selling polices for crops not eligible for benefits (i.e., specialty crops) under federal commodity price and income support programs. ***[Sec. 11011]***
Noninsured Crop Assistance Program (NAP)		
The Noninsured Crop Assistance Program (NAP) has permanent authority under Section 196 of the Federal Agriculture Improvement and Reform Act of 1996, and receives such sums as necessary in mandatory funding. Growers of crops not insurable under the crop insurance program are eligible for NAP. A payment is made to an eligible producer whose actual production is less than 50% of the established (historical) yield for the crop.	Through FY2017, makes available additional coverage for NAP at 50% to 65% of established yield and 100% of average market price. Premium for additional coverage is 5.25% times the product of the selected coverage level and value of production (acreage times yield times average market price). The premium for additional coverage is reduced by 50% for limited resource, beginning, and socially disadvantaged farmers.	Similar to Senate bill except as indicated below. ***[Sec. 11024]***
	For producers with fruit crop losses in 2012, payments associated with additional coverage are made retroactively	No comparable provision.

Current Law/Policy	Senate-Passed Farm Bill (S. 3240)	House Agric. Comm. Bill (H.R. 6083)
The payment rate is 55% of the average market price. Producers pay a fee of $250 per crop per county, or a maximum of $750 per producer per county, not to exceed $1,875 per producer. ***[7 U.S.C. 7333]***	(minus premium fees) in counties declared a disaster due to freeze or frost.	
	Eliminates NAP for crops/grasses used for grazing (to reduce overlap with livestock disaster programs in Title I—Commodity Programs), ferns, and tropical fish.	No comparable provision.
	Increases base NAP fee to $260 per crop per county, or $780 per producer per county, not to exceed $1,950 per producer. ***[Sec. 12204]***	No comparable provision.
Tree Assistance Program (TAP)		
Beginning in 2008, five new disaster programs were authorized for disasters occurring on or before 9/30/11. ***[7 U.S.C. 1531]*** One of the five is the Tree Assistance Program (TAP), which provided payments to eligible orchardists and nursery growers to cover 70% of the cost of replanting trees or nursery stock and 50% of the cost of pruning/removal following a natural disaster. Maximum TAP payments set at $100,000 per person per year.	Four programs including TAP are reauthorized with mandatory funding from the Commodity Credit Corporation for FY2012 through FY2017. TAP payment rate for replanting is reduced from 70% to 65%. **[Sec. 1501]**	Same as the Senate bill except maximum payment limit is increased to $125,000. **[Sec. 1501]**

Source: CRS Report R42552, *The 2012 Farm Bill: A Comparison of Senate-Passed S. 3240 and the House Agriculture Committee's H.R. 6083 with Current Law*.

Notes: FCIC is the Federal Crop Insurance Corporation, which is the funding mechanism for the federal crop insurance program. USDA's Risk Management Agency (RMA) administers the program. USDA's Farm Service Agency administers the Noninsured Crop Assistance Program (NAP) and the Tree Assistance Program (TAP).

End Notes

[1] Additional information on the federal crop insurance program in general is available in CRS Report R40532, *Federal Crop Insurance: Background and Issues*.

[2] Specialty crops include fruits and vegetables, tree nuts, dried fruits, and horticulture and nursery crops (including floriculture). This definition was most recently modified in the 2008 farm bill (P.L. 110-246). For more information on specialty crops and federal programs, see CRS Report R42771, *Fruits, Vegetables, and Other Specialty Crops: Selected Federal Programs*.

[3] In practice, the crop insurance company bills the farmer for the producer's portion of the premium (i.e., excluding the government subsidized portion). The company then sends the entire producer-paid premium to RMA. When a producer files a claim and the company pays an indemnity, RMA reimburses the company in full for the loss. At the end of the reinsurance year, there is an annual settlement whereby the company's proportion of any underwriting gain or loss is determined and paid.

[4] For more on the history of federal crop insurance, see http://www.rma.usda.gov/aboutrma/what/history.html. Law citations are the Federal Crop Insurance Act of 1980 (P.L. 96-365), the Federal Crop Insurance Reform Act of 1994 (P.L. 103-354), and the Agriculture Risk Protection Act (ARPA) of 2000 (P.L. 106-224).

[5] U.S. General Accounting Office, *Crop Insurance: USDA's Progress in Expanding Insurance for Specialty Crops*, GAO/RCED-99-67, April 1999, http://www.gao.gov/archive/1999/rc99067.pdf.

[6] A list and maps showing availability by county are available at http://www.rma.usda.gov/data/cropprograms.html. Specialty crops with policies include almonds, apples, avocados, bananas, blueberries, cabbage, chili peppers, 19 policies on various citrus, coffee, cranberries, dry beans, dry peas, figs, fresh market beans, fresh market sweet corn, fresh market tomatoes, grapes, green peas, macadamia nuts, mint, mustard, nursery, olives, onions, papaya, pears, pecans, peppers, pistachios, popcorn, potatoes, processing beans, pumpkins, raisins, 10 policies on stone fruit (cherries, fresh apricots, fresh freestone peaches, fresh nectarines, peaches, plums, processing apricots, processing cling peaches, processing freestone peaches, prunes), strawberries, sweet corn, sweet potatoes, table grapes, tomatoes, 13 types of fruit and nut trees, and walnuts. Crops without policies (and with acreage reported by USDA) include the following: artichokes, asparagus, blackberries, boysenberries, broccoli, cantaloupes, carrots, cauliflower, celery, cucumbers, dates, garlic, guavas, hazelnuts, honeydews, kiwi fruit, lettuce, raspberries, spinach, squash, tart cherries, and watermelons.

[7] Keith Collins, "Crop Insurance & Specialty Crops," *Crop Insurance Today*, August 2012, pp. 35-40, http://www.ag-risk.org/NCISPUBS/Today/2012/Articles/August/SpecialtyCrops-Final.pdf.

[8] Additional information on NAP is available at http://www.fsa.usda.gov/Internet/FSA_File/nap_august_2011.pdf.

[9] An indemnity is not paid if crop loss is caused by insufficient or improper applications of pest or disease control measures. For some crops such as California and Arizona navel oranges and California avocadoes, RMA has a Quarantine Endorsement Pilot Program (QEPP) that, for an additional premium, provides insurance protection when a quarantine affects all or part of the insured crop. For more information, see http://www.rma.usda.gov/policies/2011/11-qe.pdf and http://www.rma.usda.gov/handbooks/24000/2011/24250.pdf.

[10] A detailed explanation of crop insurance with examples is available in USDA, Risk Management Agency, *Crop Insurance Option for Vegetable Growers*, October 2008, http://www.rma.usda.gov/pubs/2011/vegetable_growers.pdf.

[11] One of the newest programs is for pistachios (2012) in 21 counties in California, two counties in Arizona, and one in New Mexico. The APH-Pistachio Pilot Program is a two-year production-based policy that addresses the crop's alternate-bearing production pattern.

Additional information on insurance for perennial crops and related issues is in *Report to Congress: Perennial Crops, Pecans, and the Federal Crop Insurance Program*, USDA, Federal Crop Insurance Corporation, Washington, DC, July 2, 2009, http://www.rma.usda.gov/pubs/2009/perennialcrops.pdf.

[12] The insured causes of loss include adverse weather but not price declines. For more information on the nursery program, see http://www.rma.usda.gov/pubs/rme/nursery.pdf. For additional background, see New York State Crop Insurance Education, "Nursery Crop Insurance," April 2012, http://www.agriculture.ny.gov/AP/slides/2012-13-NYS- Nursery-Crop-Insurance.pdf.

[13] Nursery insurance covers all field-grown and container nursery plants that are on the eligible plant list (specifying more than 20,000 plants) and meet other requirements including appropriate growing medium and production practices. More information is available in USDA, Risk Management Agency, *Crop Insurance for Nurseries*, Washington, DC, October 2008, http://www.rma.usda.gov/pubs/2011/nursery. pdfhttp://www.rma.usda.gov/pubs/2011/nursery.pdf; and James B. Johnson and John Hewlett, *Nursery Crop Insurance in Wyoming*, University of Wyoming, Agricultural Marketing Policy Paper No. 15, September 2006, http://www.ampc.montana.edu/ policypaper/policy15.pdf.

[14] For more information, see USDA fact sheet at http://www.rma.usda.gov/pubs/rme/agr-lite.pdf.

[15] USDA/Risk Management Agency, *Quarantine Endorsement Pilot Program for California Avocado and Citrus*, Commodity Insurance Fact Sheet, March 2012, http://www.rma.usda.gov/fields/ca_rso/2012/caquarantine.pdf.

[16] The calculation is 7.25 million acres insured divided by 13.77 million acres. The denominator is from USDA's *2007 Census of Agriculture*, Specialty Crops, Volume 2, Subject Series, Part 8; it includes acreage for nursery and floriculture not included in the RMA figure for insured acres. http://www.agcensus.usda.gov/Publications/2007/Online_Highlights/Specialty_Crops/speccrop.pdf.

[17] *Report to Congress: Specialty Crop Report*, USDA, Federal Crop Insurance Corporation, Washington, DC, November 2010, p. 14, http://www.rma.usda.gov/pubs/2010/specialtycrop.pdf.

[18] The subsidy ranges from 67% for 50% yield coverage of (i.e., yield loss must greater than 50%, and more than 55% of expected market price) and to 38% for yield coverage of 80% (i.e., yield loss must greater than 20%).

[19] Crop insurance premium subsidies totaling $148 million for other items such as forage, pasture, and tobacco are excluded here because they are not eligible for commodity program subsidies. The list of specialty crops employed here is the same as the one used for the 2007 Census of Agriculture. These commodities include fruits, tree nuts, berries, vegetables, potatoes, melons, pulse crops, nursery, greenhouse, floriculture, and other specialty crops. Crop insurance premium subsidy data from this section have been summarized from USDA/RMA data published at http://www.rma.usda.gov/data/sob.html.

[20] Total crop value in 2011 for these two broad categories is estimated at $196 billion, based on a CRS compilation of various USDA data. Specialty crops account for $61 billion or 31% of the total, and program crops account for $135 billion or 69% of the total. The estimates are based on estimates of 2011 cash receipts of major program crops, fruits, nuts, vegetables, and melons by USDA's Economic Research Service, 2011 estimates of crop value for other program crops by USDA/National Agricultural Statistics Service, and an estimate of horticulture crop value from Table 37 in the 2007 Census of Agriculture, http://www.agcensus.usda.gov/Publications/2007/Full_Report/Volume_1,_Chapter_1_US/st99_1_037_037.pdf.

[21] Ibid., p. 8. Also, organic acreage under insurance is summarized by state and crop on pages 20-24.

[22] Organic agriculture can include all types of commodities, including grains, fruits and vegetables, and animal products.

[23] For a detailed report on insurance for organic crops, see *Organic Crops and the Federal Crop Insurance Program*, USDA, Federal Crop Insurance Corporation, Washington, DC, January 28, 2010, http://www.rma.usda.gov/pubs/2010/ organics.pdf. The report was required by the 2008 farm bill (P.L. 110-246).

[24] *Risk Management Agency Fact Sheet—Organic Farming Practices*, Program Aid 1912, USDA Risk Management Agency, Washington, DC, January 2011, http://www.rma.usda.gov/pubs/rme/organics.pdf. Additional information is available at http://www.rma.usda.gov/news/currentissues/organics/.

[25] This section is summarized in part from Keith Collins, "Crop Insurance & Specialty Crops," *Crop Insurance Today*, August 2012, http://www.ag-risk.org/NCISPUBS/Today/2012/Articles/August/SpecialtyCrops-Final.pdf.

[26] Adverse selection occurs when producers purchase insurance if they expect indemnities to exceed the premium, or do not purchase insurance if they expect indemnities to be less than the premium. Moral hazard is another typical issue for all forms of insurance. It occurs when an insured producer alters his or her behavior to increase the chances of collecting an indemnity. An insurance deductible, whereby the initial loss is paid by the insured, is often employed to reduce costs associated with moral hazard.

In: Specialty Crops
Editor: Katelin Fransen
ISBN: 978-1-62618-054-3

Chapter 3

REPORT TO CONGRESS: SPECIALTY CROP REPORT*

United States Department of Agriculture

EXECUTIVE SUMMARY

The 2009 Specialty Crop Report is submitted in response to Section 508 (a)(6)(B) of the Federal Crop Insurance Act (Act), as amended, requiring that the Federal Crop Insurance Corporation (FCIC) report to Congress on the progress and expected timetable for expanding crop insurance coverage under the Act to new and specialty crops. As indicated by the requirements of the law, the report is a useful way to obtain a quick overview of the processes and timelines the Risk Management Agency (RMA) must follow in order to make new and specialty crop insurance products available to producers.

This report complements *Participation in the Federal Crop Insurance Program* (Participation Report) submitted in accordance with Section 508(a)(7) of the Act. The Participation Report was prepared to assist the FCIC Board of Directors (Board) in establishing development priorities for the future. The Participation Report also outlined the progress that the FCIC has made since a comprehensive analysis was completed of the crop insurance market. As detailed in the Participation Report, the Board questioned whether FCIC may be nearing a saturation point where the potential for cost effectively

* This is an edited, reformatted and augmented version of the United States Department of Agriculture, Federal Crop Insurance Corporation, dated November 2010.

adding new crop programs is limited. However, as new industries emerge as legitimate agricultural enterprises, RMA continually monitors the need for additional coverage options for these developing industries.

The growth of specialty crop programs can be measured by the frequency with which new and/or revised product offerings for specialty crops are considered by the Board. During calendar year 2009 alone, the Board was presented with new or modified products for the following specialty commodities: apiculture, avocados, citrus, dry peas & lentils, fresh market beans, strawberries, sugar beets, sugarcane, and tobacco.

Each year RMA responds to a number of commodity organizations and industry representatives who either desire a new Federal crop insurance product or an enhancement to the existing coverage options. Enhancement requests often focus on revenue coverage, which was first made available to the major commodities (e.g., corn, soybeans) beginning in the mid-1990's and bases coverage on values obtained from commodity futures exchanges. These major commodities have since experienced a rapid shift from yield based to revenue coverage over the past decade which is evidence of the demand for revenue coverage among agricultural producers. However, there is no such comparable price discovery mechanism for specialty crops, and therefore development of a viable revenue insurance product is much more difficult. Nevertheless, RMA remains committed to exploring all options for possibly providing revenue coverage to specialty crops.

For example, the Board recently approved a new revenue coverage product, Actual Revenue History (ARH), for cherries, navel oranges and strawberries. Given its design, the ARH product is most appropriate for fresh fruit and perishable commodities. Conversely, the applicability of ARH beyond this niche can be problematic because basic assumptions of the ARH design are violated.

In considering new crop programs designed to benefit specialty crop producers, RMA and the Board are faced with the challenge of allocating limited resources across competing interests. This task has become more challenging given the requirements of Office of Management and Budget Memorandum, "Budget Discipline for Agency Administrative Actions", dated May 23, 2005, commonly known as PAYGO requirements, and RMA's relatively static operational budget. Further, expanding existing specialty crop programs to additional growing areas also requires meeting PAYGO requirements because these expansions are discretionary actions. Thus, without a corresponding reduction, or savings, in another part of the crop insurance program or in another USDA program, new and/or expanded programs cannot be implemented because funding is not available.

Low participation, high loss ratios, and program design limitations learned through the pilot program process have resulted in the Board terminating several crop insurance programs. Program terminations over the past several years include: crambe, processing cucumbers, fresh market beans, raspberries, blackberries, sweet potatoes, and winter squash.

SPECIALTY CROP DEFINITION

The term "specialty crops" is a broad term that is defined differently throughout government. Historically, RMA has reported program participation levels based on the following ten major crops: barley, corn, grain sorghum, peanuts, potatoes, rice, soybeans, tobacco, upland cotton, and wheat. All other crops have often been referred to as specialty crops; however, this classification is no longer consistent with legislation.

The Agricultural Economic Assistance Act of 2001 defined specialty crops as "any agricultural crop, except wheat, feed grains, oilseeds, cotton, rice, peanuts and tobacco."

This broad definition led to various interpretations by Federal agencies and state governments. The definition of specialty crops was refined in the Specialty Crops Competitiveness Act of 2004 (SCCA) as "fruits and vegetables, tree nuts, dried fruits and nursery crops (including floriculture)". This definition was further refined in the Food, Conservation and Energy Act of 2008 by the inclusion of "horticulture and" in front of the word "nursery" in the SCCA definition.

For the purpose of this report, the more narrow definition from the Food, Conservation and Energy Act of 2008 is used. The Adjusted Gross Revenue (AGR) and AGR-Lite plans of insurance also provide specialty crop coverage. Although it is possible to insure virtually any crop under these two insurance products, the target niche of the AGR portfolio, as well as the bulk of current participation, is within the specialty crop segment.

Appendix A Tables 1 and 2 contain a listing of specialty crops as provided by USDA. These tables also depict each crop's status within the Federal crop insurance program and denote individual crop coverage separately from whole farm AGR coverage. This list is not intended to be all-inclusive, but rather is intended to give examples of the most common crops.

NEW PRODUCT DEVELOPMENT PROCESS

The Act has established two pathways for the development of new crop insurance programs; internal products developed under contract and external products submitted through the procedures in section 508(h) of the Act. Once the program materials are developed, both the internal and external products undergo the same approval process. The approval process includes a great deal of oversight, is intensive and can take up to a year to complete.

Section 522(e)(4) of the Act prohibits new product research and development by the FCIC stating: "…on and after October 1, 2000, the Corporation shall not conduct research and development for any new policy for an agricultural commodity offered under this subtitle." Instead Section 522(c) of the Act provides contracting authority stating: "the Corporation may enter into contracts to carry out research and development."

Section 508(h) of the Act provides a mechanism whereby a private sector entity can propose an insurance plan to be added to the FCIC portfolio of products stating: "…a person (including an approved insurance provider, a college or university, a cooperative or trade association, or any other person) may prepare for submission or propose to the Board other crop insurance policies and provisions of policies; and rates of premiums for multiple peril crop insurance…"

The Food, Conservation, and Energy Act of 2008 provided for an enhancement to the 508(h) process whereby the Board can choose to approve an advance partial payment for the development of a new product. Section 522(b)(2)(A) states "...the Board may approve the request of an applicant for advance payment of a portion of reasonable research and development costs prior to submission and approval of the policy by the Board under section 508(h)."

508(h) products approved by the Board are generally eligible for federal subsidies; and the private submitter is eligible for reimbursement of the development cost. If the private submitter continues to provide support for the product, they are eligible to bill for maintenance costs during an initial four year period. They can, at any point, relinquish support of the product thus surrendering any additional year's maintenance fees.

All new products must be approved by the Board through a rigorous set of procedures. The initial step in the approval process is to submit complete program materials – policies, actuarial materials, etc. – to the Board. The Board then must choose whether or not to move the product forward for an independent external review. The external reviewers are selected by the Board from a pre-approved pool of applicants with professional knowledge and experience in crop insurance principles and actuarial science. Generally, five reviewers are selected for each proposed product.

Each reviewer will typically highlight areas of potential weakness or suggested improvement within the product. It is incumbent upon RMA or the private submitter to address the issues identified by the external reviews. Once these issues have been addressed, the product is eligible to return before the Board for approval. In certain instances the issues raised by the external reviewers may require substantial program changes, in which case the product must be reviewed for a second time.

The speed at which a product progresses through the approval and implementation process is contingent upon many factors. The most critical factor, however, is the quality and thoroughness of the submission package presented to the Board, and the responsiveness of the submitter to issues raised by the Board and the reviewers.

Approved products are generally implemented as pilot programs and are not published in the Federal Register. The Act provides pilot authority as a mechanism to test either new insurance designs or to cover crops which were previously uninsurable in the program. Section 523(a)(1) of the Act states: "...the Corporation may conduct a pilot program submitted to and approved by the Board under section 508(h), or that is developed under subsection (b) or section 522, to evaluate whether a proposal or new risk management tool tested by the pilot program is suitable for the marketplace and addresses the needs of producers of agricultural commodities."

Legislation also prohibits FCIC from offering a pilot program that would compete with coverage that is already generally available in the private sector. The Act states: "…the Corporation shall not conduct any pilot program that provides insurance protection against a risk if insurance protection against the risk is generally available from private companies."

Pilot programs are sometimes limited in scope so that during the pilot phase the program can be tested for effectiveness while limiting exposure. Pilot programs are more adaptable than permanent programs because changes can be made without going through the full regulatory process which involves a proposed and final rule. If proposed pilot changes are material in nature they require formal Board action.

Pilot programs generally operate for four years, but this period could be longer in order to allow maximum flexibly to test and enhance a new product. During this pilot period the program is monitored by RMA to assess whether the program is operating as intended and make adjustments as needed to assure program viability. At the conclusion of the pilot period the program is evaluated by an independent third party contractor. The review as well as RMA's assessment is offered to the Board. At that time the Board has full latitude to: a) extend the pilot program for additional years; b) authorize conversion of the program to regulatory; or c) terminate the program.

Internal Development of Products for Specialty Crops

In recent years several notable specialty crop programs have been developed through the internal procedure and authority and approved by the Board for implementation.

Quarantine Endorsement Pilot Program

Beginning with crop year 2011, quarantine coverage is available to citrus and avocado growers in the state of California. This product originated from the concerns of various commodity organizations who have expressed interest in an insurance program to cover losses due to quarantines imposed by a duly authorized regulatory body, such as the Animal and Plant Health Inspection Service (APHIS).

The Act limits insurance coverage to loss of production attributable to natural perils or for revenue plans, adverse price movement. Under the statute, RMA has a narrow authority to cover production loss due to quarantine. Specifically, only quarantine coverage due to a natural pest or disease is eligible for consideration. The core quarantine area is defined as the area infected or having a high likelihood to become infected, compartmentalized by the authorized regulatory authority and supported through science.

Production losses attributed to pest infestation, disease, or a declared qualifying quarantine which limits or prohibits harvest, and the ability to transport the insured crop, embody the coverage offered under this endorsement. RMA cannot cover economic losses due to missed marketing opportunities caused by quarantine for crop production not damaged. Additionally, to be eligible for an indemnity a grower must follow best management practices to preserve the crop during the quarantine period. The coverage will not compensate the grower for these potential additional costs.

Actual Revenue History (ARH) Pilot

The ARH insurance design has many parallels to the widely available Actual Production History (APH) Plan of Insurance, with the primary difference being that instead of insuring historical yields, this plan insures historical revenues. Beginning in crop year 2009 an ARH program for cherries was approved by the Board. Subsequently, ARH programs for citrus (navel oranges) and strawberries were approved with coverage beginning in crop year 2011 and 2012, respectively.

Providing revenue coverage to crops that lack a public, centralized price discovery mechanism such as a commodity futures exchange is exceedingly challenging. Current FCIC revenue coverages (Revenue Assurance (RA) and Crop Revenue Coverage (CRC)) use a futures price to provide intra-season price protection. A predetermined methodology is used to develop the projected harvest price prior to planting, based on the harvest futures contract (December corn futures on the Chicago Board of Trade). This same contract is later used to determine the actual harvest price for the commodity during the harvest period. These prices are used to calculate a grower guarantee and determine any potential grower indemnity.

ARH-Cherry Participation

The ARH-Cherry pilot program was implemented for crop year 2009. Participation in the first year program was strong with almost 1,750 polices sold; this represents a slight increase to the policy count from the 2008 crop year in the now terminated pilot dollar plan of insurance. Over 50,000 acres of cherries were insured for crop year 2009. Washington leads the way with approximately 23, 000 acres insured followed by California at just over 20,000 acres. These two largest cherry producing states had participation scores of 70% and 75%, respectively.

ARH is different than this traditional FCIC revenue coverage in that it guarantees an average of historical grower revenues. Like other revenue coverage, the ARH plan does not intend to provide inter-season price protection because doing so will allow for adverse selection. However, for perishable crops, stocks are not carried over across growing seasons so, absent significant supply or demand shocks, expected prices remain relatively stable across years. In this environment, an average of historical prices (or grower revenues) is a reasonable predictor of future prices (or grower revenues). The ARH plan provides for an adjustment mechanism in the event that a significant market change or "shock" is observed prior to the sales closing date. It is believed that this adjustment will be relatively rare for fresh market, perishable crops.

The ARH niche may be best determined by examining all available pricing information. In general the initial ARH commodities share common pricing traits:

- The overall availability of price information is thin, thus limiting a grower's ability to select against the offer.
- Growers do not possess additional price information through a contracting arrangement or some other micro level outlet.
- Grower harvest price will likely be a function of the overall production of the region, often inversely correlated.
- The short term average of historical prices is a reasonable predictor for future prices.

For fresh market, perishable crops, an average of historical revenues can provide a reasonable estimate of expected future revenues, and therefore serve as the basis for the insurance guarantee. This is the conceptual basis for the ARH program design. However, for other crops (e.g., storable commodities, etc.) this assumption is less tenable.

REGIONAL OFFICES

Educational efforts are an important component to the delivery of crop insurance and are especially important for first year programs. From a grower perspective, the RMA Regional Offices are the face of the organization and play a critical role in these efforts. During the ARH-Cherry pilot program implementation these offices sponsored grower training sessions to promote education and assist growers through this transition year.

PRIVATE SECTOR INITIATIVES FOR SPECIALTY CROPS

Section 508(h) of the Act authorizes the FCIC to reimburse private entities for research, development, and maintenance costs if they develop an insurance program that is approved by the Board. In recent years several new specialty crop programs have originated utilizing this developmental pathway.

Processing Pumpkins

- Approved for the 2009 crop year with coverage available in 11 counties in Illinois.
- Provides APH based coverage for irrigated and non-irrigated processing pumpkins.
- Pumpkins must be grown under contract with a processor.
- The program potentially covers 95 percent of all processing pumpkin production.

Apiculture

- Approved for the 2009 crop year with coverage available in Alabama, Arizona, California, Colorado, Florida, Georgia, Idaho, Kansas,

Minnesota, Missouri, Montana, Nebraska, New Mexico, New York, North Carolina, North Dakota, Oklahoma, Oregon, Pennsylvania, South Carolina, South Dakota, Texas, Utah, Virginia, and Wyoming.

- A group risk program that insures honey production.
- Uses the same indexes, technology, methodology, and rating methods that are currently being used for the Pasture, Rangeland, and Forage (PRF) Rainfall Index and Vegetation Index pilot programs.
  - The Rainfall Index is based on weather data collected and maintained by National Oceanic and Atmospheric Administration's (NOAA) Climate Prediction Center. The index reflects how much precipitation is received relative to the long-term average for a specified area and timeframe.
  - The Vegetation Index is based on the U.S. Geological Survey's Earth Resources Observation and Science normalized difference vegetation index data derived from satellites observing long-term changes in greenness of vegetation.
- Honey production was demonstrated to be directly dependent upon forage growth/productivity, which is the basis for the PRF pilot programs.

Plantain and Banana

- Approved for the 2009 crop year with coverage available in Puerto Rico only.
- Only covers losses due to damage from a tropical cyclone classified by the National Hurricane Center in Miami as a hurricane.
- The program potentially covers approximately 98 percent of plantain and banana producers.

Sugarcane

- Approved for the 2009 crop year with coverage available in Louisiana.
- Group risk crop insurance plan that does not rely on an individual's actual production to determine whether an indemnity is due.
- Designed to provide insurance benefits comparable to other group risk crop insurance programs administered by FCIC.

- Uses yield and price data from the National Agricultural Statistics Service (NASS) and planted acres data from the Farm Service Agency (FSA).
- Protects against reduction in sugarcane yields occurring resulting from unavoidable natural events.

Fresh Market Beans

- Approved for the 2011 crop year with coverage available in New York, North Carolina and Virginia.
- Provides Actual Production History based coverage.
- Eligible producers must provide four years of fresh market bean production records.

PLANNED SPECIALTY CROP RESEARCH AND DEVELOPMENT

The FCIC is required to enter into contracts for research and development rather than directly performing those duties. Prior to entering into a contract for these research and development initiatives, RMA must consult relevant commodity organizations. These organizations provide RMA information with regard to producer interest in crop insurance programs as well as comments on policy structure or other risk management options available. If sufficient interest is indicated, a contract may be awarded to begin research and development for a new commodity. The development may be preceded by a feasibility report.

In the upcoming year, RMA plans to assess the insurance potential for a number of crops. Most commodity organizations seeking a FCIC pilot program for the first time request the APH plan of insurance because this offers the grower individual yield based coverage. However, for a commodity to be insurable under this plan, certain basic characteristics must be demonstrated. The commodity must have established cultivars, reasonably well defined farming practices, developed markets and known perils. Establishing a commodity risk profile involves research and data collection in these areas. Typically USDA agencies, universities, commodity organizations and private companies provide this type of information and research.

Biofuel Crops

RMA will explore several underdeveloped crop industries that hold some potential for future development, especially in the biofuel arena. Insurance coverage availability may be an important factor to assist in the development of these industries. RMA continues to monitor the emergence of industries related to cellulosic ethanol production and will commission a research report in the near future to examine these industries.

Pistachio Nuts

RMA has initiated development of an APH program to insure pistachio nuts. A feasibility study on developing crop insurance for pistachios conducted in 1995 concluded that pistachios growers had little interest in crop insurance. While a privately available insurance product was available at one time, participation in the program had greatly declined. Instead, pistachio growers recently indicated a strong interest in a FCIC multiple peril crop insurance program. Many pistachios growers are diversified nuts growers who use FCIC products for other commodities. RMA conducted a Notice of Intent to Research Feasibility/Develop Insurance for Pistachios during calendar year 2009 and received several favorable comments from the industry.

Pistachio nuts, unlike some specialty commodities, do not lack for a market. In 2008, there were 118,000 bearing acres of pistachios, production was 139,000 tons, and the pistachio utilized crop value was 550 million dollars. The majority of production occurs in California where pistachio nuts rank as the 16th largest cash value crop for the state and the fifth largest fruit/nut crop.

Revenue Coverage

Research and development of revenue coverage remains a high priority for RMA in the upcoming years. The ARH plan of insurance is the newest design to provide growers with revenue coverage.

RMA intends to fully utilize the pilot phase of the current ARH programs to monitor performance. The pilot phase is a valuable period where RMA has the opportunity to evaluate new program functionality and ensure that the program is operating as intended. Part of the process allows for an assessment of whether grower needs are being met in a fair and appropriate manner. This evaluation process allows for RMA to make minor program adjustments while the scope is still relatively minor and the policy is not yet codified.

RMA is aware of several commodity groups that have interest in the ARH design beyond the crop segment where the program is currently targeted--fresh market, perishable crops. For these targeted perishable crops it is believed that an average of historical revenues can provide a reasonable estimate of expected future revenues, and therefore is the basis for the insurance guarantee. This is the conceptual basis for the ARH program design. However, for other crops (e.g., storable commodities, etc.) this assumption is less tenable. If the price is reasonably predictable and this projected price does not track well with a short term moving average, then the ARH plan is likely a poor fit.

RMA is evaluating several of these crops on an individual basis to best determine the most appropriate crop list for the ARH niche.

Named Peril Weather Insurance

RMA recently developed a new insurance design called Named Peril Weather Insurance (NPWI), which offers growers an innovative and unique form of coverage based on specific weather events. This product would allow individual producers to "select" up to five weather perils and perilous windows (or periods).

Both the suite of weather perils insured and the monitoring location (weather stations) are elected by individual producers. At sign-up, the insured defines critical time periods (start and end dates) and weather conditions (amounts of precipitation, temperatures, or degree-days) to be insured. Input about the potential extent of losses from each peril (i.e., a loss function) is also supplied by the producer.

This design has many attractive features, one being that many different crops could potentially be covered under the same design. This is an important factor in terms of development and maintenance costs relative to the total number of growers and crops potentially impacted. Each crop initially considered under this design individually comprises a small percentage of total farm receipts nationally. However, collectively these twenty plus crops could impact up to one- half billion in total farm receipts.

During the final stages of development, RMA became aware that several private sector companies were offering similar coverage. Under Section 523 (a) (2) of the Act "…the Corporation shall not conduct any pilot program that provides insurance protection against a risk if insurance protection against the risk is generally available from private companies." This prohibition is applicable only to products developed by RMA.

These private sector companies, which traditionally offer weather based coverage options for other industries and/or events, are still adapting their products to meet the needs of agricultural producers; nonetheless this coverage was determined to be generally available. At the request of several private companies and in compliance with the statute, NPWI was tabled until such a time that it can be determined whether the private products are meeting the needs of specialty crop growers. RMA will continue to monitor the status of the privately available products to assess whether the NPWI product development can continue.

Crop Program Expansion

Federal crop insurance is administered at the county level. In some cases a regulatory crop insurance program may already exist, however coverage may not be available in a particular county. The expansion of crop programs to additional counties may become necessary, for example, when acreage shifts to a new growing region. Generally, crop program expansion is an annual event that takes place before the contract change date for the particular crop. For most crops, specialty crops included, the key factors in determining whether a crop program will be established in the county are:

- significant grower interest in having insurance protection;
- the economic significance of the crop;
- sufficient data on which to establish a program;
- acceptable risk assessment; and
- available offsets to satisfy the PAYGO requirements.

Program expansion requests are most normally initiated by producers at the local level and channeled through RMA's Regional Offices. RMA gathers production data and information to evaluate the requirement for actuarial sufficiency. Additional cost estimates are prepared and all information is presented to the RMA Administrator who has authority to determine county crop program expansions.

While crop insurance provides producers the option to obtain more protection than through the Non-insured crop disaster Assistance Program (NAP), and reduces their reliance on *ad hoc* disaster assistance, expansion also increases FCIC's discretionary and mandatory spending. Expansions are only approved after the appropriate offsets have been determined in compliance with PAYGO requirements. Until a county expansion is approved, growers also have the option of applying for a written agreement on most regulatory programs.

IMPROVING CURRENT SPECIALTY CROP PROGRAMS

The Participation Report suggests that continued refinement of current FCIC programs to address producer needs is a sound use of RMA resources and may well increase market penetration for these crops that already have crop insurance coverage. These refinements occur both as part of the product development process as well as within the regulatory process. RMA converted the California Avocado Program to an APH program for crop year 2010 and is in the process of converting the Chile Pepper Dollar Plan of Insurance to an

APH program. Both conversions occurred as the result of a pilot program evaluation and subsequent Board action.

The regulatory process allows for program refinement and improvement for all crop programs, including specialty crops. Two high profile specialty crops, wine grapes and table grapes, were recently revised through the regulatory process with new policies going final rule for crop year 2010. Wine grape production has increased throughout the country in recent years and many varietals previously grown only in California are now produced around the country. A key feature of the new policy is to offer varietal coverage in other states similar to what was available in California under the previous policy.

Nursery

Nursery, by the definition in the Act is a specialty crop, and is currently entering the regulatory refinement process. Nursery is a far-ranging and diverse industry which challenges traditional crop insurance designs. Thus, the Federal crop insurance nursery program utilizes a unique product design. Instead of insuring a percentage of average historical production or a percentage of average historical revenue, typical of most FCIC products, the nursery program is an asset- based form of insurance coverage. This coverage does not include protection due to plant price fluctuation. Nursery crop insurance is available in all States to all persons operating nurseries that meet certain criteria. In crop year 2008, nationwide coverage was approximately 4 billion dollars with Florida as the number one state at just over 1.25 billion dollars in liability.

Nursery crop insurance is available in all States to all persons operating nurseries that meet certain criteria. Field grown and container grown plants are provided coverage against damage resulting from insurable causes of loss occurring during the insurance period. The insurable causes of loss are similar to those covered under multiple peril crop insurance policies which are most generally adverse weather events such as freeze and hurricane.

The underwriting process for nursery is intensive and detail oriented. For the 2010 crop year, there are approximately 25,500 insurable plants as contained on the Eligible Plant List and Plant Price Schedule (EPLPPS) which is published by FCIC. At application, each insured nursery grower must submit a Plant Inventory Value Report (PIVR), the nursery wholesale catalog or price list, and a Crop Inventory Value Report (CIVR) to the insurance provider. These reports are used in the guarantee determination process.

RMA is currently soliciting for proposals for an evaluation of the nursery program. The solicitation for proposals closes in early March, 2010.

Adjusted Gross Revenue (AGR)

AGR first debuted in the FCIC portfolio of products back in crop year 1999. AGR is whole farm revenue coverage with the guarantee derived primarily from historical farm revenue reported on producer tax forms. The plan provides producers with protection against low revenue from natural causes (production loss) and also due to market fluctuations (price decline). AGR insures historical income from agricultural commodities, as well as income from animals, animal products, and aquaculture species reared in a controlled environment. In general AGR is targeted at specialty crops and/or small crops which do not have individual crop coverage options.

There was rapid expansion of the AGR program during the first half of the decade and considerable positive momentum surrounding the plan. In addition to the AGR pilot plan there was a 508(h) submission- AGR-Lite which began in crop year 2003. AGR-Lite is very closely modeled off the AGR plan, loosening only a few underwriting requirements, and lessening the maximum amount of coverage. Participation as measured in policy count peaked in crop year 2003 with just over 1000 policies earning premium. In recent years interest in the program has diminished and participation has faded to approximately 850 policies earning premium for crop year 2008 and slightly fewer than 800 for crop year 2009.

RMA commissioned an independent study to determine if there were reasonable improvements which would increase interest in the plan. There study concluded there were many challenges facing AGR which include:

- Producer reservations to submit tax records for insurance;
- Complex and paperwork intensive;
- AGR provides a relatively low level of coverage and is often viewed as catastrophic coverage;
- Lack of interest in whole farm coverage; and
- Delay with potential indemnities.

Many of these suggestions were difficult to implement without initiating a fundamental change to the AGR methodology. While RMA continues to examine possible improvements to the program; it is appropriate to acknowledge AGR as a niche product. As a niche product it is unfair to assume AGR can provide a solution to all crops currently lacking coverage. However, when properly understood AGR brings value to a minor segment of the current population of insureds.

SUMMARY

Managing growth, expansion and refinements to specialty crop programs within a tight operational budget will be a growing challenge for the Board and the leadership of RMA in the coming years. However, RMA has positioned itself well by developing broad concepts such as ARH and NPWI that have the potential to impact multiple crops using the same general concept.

As a benchmark to establish current participation rates for specialty crops, please see the attached RMA/NASS Specialty Crop Comparisons Tables and Maps. RMA acres are calculated using both CAT and buy-up level participation. The NASS acres were used as the complete population of cropped acres. In rare instances, NASS data were not published for a particular crop/state combination and were filled with 2007 agricultural census data. In other rare instances RMA acres exceeded NASS acres and the participation rate was capped at 100%. At the aggregate level this is only an issue for the California raisin crop where acreage reporting is difficult because raisins are an end use of the larger grape crop.

The first two pages display tables with aggregate information by crop and by state. The pages following contain a series of national maps for each crop which display state level participation percentages. The final pages detail the acreage amounts on a crop/state basis. Overall participation in specialty crops programs is quite good at 75%. This compares favorably to the participation levels for the major program crops of 83%. Important fruit, nut and vegetable states California (71%), Florida (91%), and Washington (68%) each score well. In addition, the north central region of the country where many of the pulse crops are grown; Minnesota (84%), Montana (83%), North Dakota (96%) and South Dakota (79%) all boast high participation rates.

FCIC Program Growth for Specialty Crops
Liability in Billions of Dollars by Year

	2000	2001	2002	2003	2004	2005	2006	2007	2008	2009
AGR	0.009	0.189	0.245	0.322	0.307	0.308	0.354	0.326	0.325	0.400
Vegetables	1.611	1.621	1.866	1.904	1.953	1.969	2.048	2.048	2.430	2.819
Nursery	2.357	2.599	3.006	3.283	3.598	3.888	3.674	4.010	4.037	3.187
Fruits/Trees/Nuts	3.452	3.642	3.952	3.924	4.052	4.528	5.130	5.067	5.337	6.256
Summary	7.429	8.052	9.069	9.433	9.910	10.693	11.206	11.451	12.129	12.662

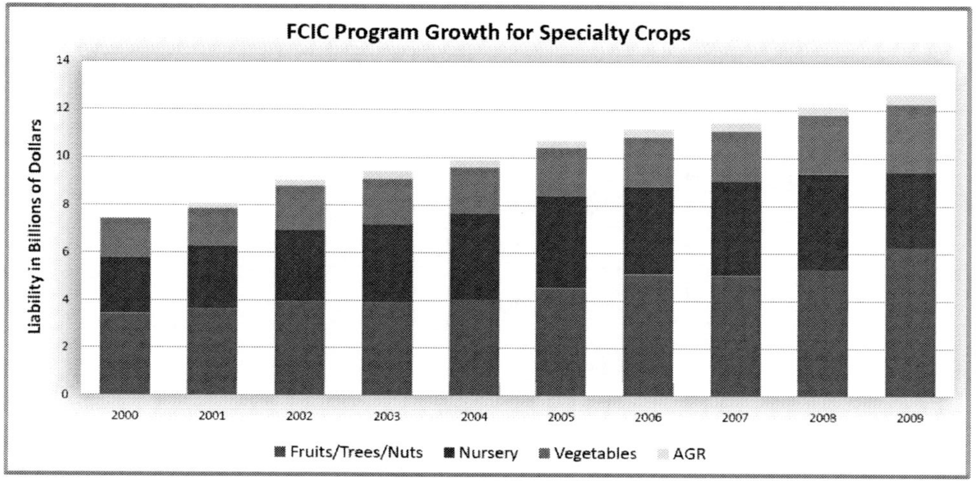

APPENDIX A

Table 1. List of Plants Commonly Considered Fruits and Tree Nuts

Almond*	Grape (including raisin)*
Apple*	Guava**
Apricot*	Kiwi**
Avocado*	Litchi**
Banana*	Macadamia*
Blackberry**	Mango**
Blueberry*	Nectarine*
Breadfruit	Olive**
Cacao	Papaya*
Cashew	Passion fruit**
Citrus*	Peach*
Cherimoya**	Pear**
Cherry*	Pecan*
Chestnut (for nuts)**	Persimmon**
Coconut	Pineapple
Coffee*	Pistachio**
Cranberry*	Plum (including prune)*
Currant	Pomegranate**
Date**	Quince**
Feijou	Raspberry**
Fig*	Strawberry***
Filbert (hazelnut)**	Suriname cherry
Gooseberry**	Walnut*

* Individual Crop Insurance Plan.
** AGR Crop Insurance Plan.
*** Individual Crop Insurance Plan Pending Implementation.

Table 2. A List of Plants Commonly Considered Vegetables

Artichoke**	Mustard and other greens**
Asparagus**	Okra**
Bean* Snap or green Lima Dry, edible	Pea* Garden, English or edible pod
Beet, table**	Onion*

Broccoli (including broccoli raab)**	Opuntia
Brussels sprouts**	Parsley**
Cabbage (including Chinese)**	Parsnip**
Carrot**	Pepper*
Cauliflower**	Potato*
Celeriac**	Pumpkin*
Celery**	Radish (all types)**
Chive	Rhubarb**
Collards (including kale)**	Rutabaga**
Cucumber**	Salsify
Eggplant**	Spinach**
Endive**	Squash (summer and winter)**
Garlic**	Sweet corn*
Horseradish**	Sweet potato**
Kohlrabi**	Swiss chard**
Leek**	Taro
Lettuce**	Tomato (including tomatillo)*
Melon (all types)**	Turnip**
Mushroom (cultivated)**	Watermelon**

* Individual Crop Insurance Plan.

** AGR Crop Insurance Plan.

RMA/NASS Specialty Crop Comparison

Obs	Crop	RMA	NASS	Percent
1	Almonds	478,380	680,000	70%
2	Apples	236,756	348,200	68%
3	Apricots	7,141	12,380	58%
4	Avocados	37,781	72,500	52%
5	Blueberries	44,055	81,536	54%
6	Cabbage	13,681	70,200	19%
7	Cherries	49,384	81,910	60%
8	Chile Peppers	3,771	15,800	24%
9	Citrus Fruit	786,286	857,041	92%
10	Citrus Trees	21,144	27,300	77%
11	Cranberries	31,616	38,500	82%
12	Dry Beans	1,276,208	1,501,511	85%
13	Dry Peas	999,132	1,171,000	85%

(Continued)

Obs	Crop	RMA	NASS	Percent
14	Figs	6,002	9,400	64%
15	Grapes	567,492	838,310	68%
16	Green Peas	172,786	224,380	77%
17	Macadamia Nuts	11,837	15,000	79%
18	Nectarines	23,009	32,300	71%
19	Onions	80,040	160,971	50%
20	Peaches	84,354	122,609	69%
21	Pears	35,104	56,000	63%
22	Pecans	156,973	279,450	56%
23	Peppers	9,204	19,000	48%
24	Plums	21,795	29,500	74%
25	Potatoes	834,062	1,065,350	78%
26	Processing Beans	102,542	209,324	49%
27	Prunes	61,752	64,000	96%
28	Pumpkins	6,188	13,100	47%
29	Raisins	249,160	249,160	100%
30	Sweet Corn	319,746	577,795	55%
31	Table Grapes	84,691	97,000	87%
32	Tomatoes	326,893	409,250	80%
33	Tropical Fruit	3,784	8,780	43%
34	Walnuts	110,071	223,000	49%
	All Crops	*7,252,820*	*9,661,557*	*75%*

State	RMA	NASS	Percent
Alabama	6,294	18,320	34%
Alaska	18	800	2%
Arizona	27,087	47,950	56%
Arkansas	1,220	3,000	41%
California	2,163,773	3,033,460	71%
Colorado	101,957	132,270	77%
Connecticut	1,866	7,100	26%
Delaware	20,163	23,812	85%
Florida	628,734	694,041	91%
Georgia	126,298	163,750	77%
Hawaii	15,621	23,780	66%
Idaho	403,216	493,605	82%
Illinois	26,337	54,297	49%

State	RMA	NASS	Percent
Indiana	7,124	25,500	28%
Iowa	4,178	5,550	75%
Kansas	7,568	11,321	67%
Kentucky	56	380	15%
Louisiana	76	240	32%
Maine	69,426	84,100	83%
Maryland	10,002	24,855	40%
Massachusetts	16,139	26,130	62%
Michigan	263,749	363,050	73%
Minnesota	344,184	410,600	84%
Mississippi	562	13,850	4%
Missouri	8,376	12,300	68%
Montana	291,459	350,810	83%
Nebraska	137,502	154,500	89%
Nevada	5,075	8,400	60%
New Hampshire	1,367	3,800	36%
New Jersey	15,386	33,650	46%
New Mexico	39,239	74,602	53%
New York	145,035	206,176	70%
North Carolina	25,253	47,866	53%
North Dakota	1,297,512	1,357,000	96%
Ohio	6,295	39,700	16%
Oklahoma	5,325	22,700	23%
Oregon	95,514	184,350	52%
Pennsylvania	43,420	79,300	55%
Rhode Island	520	1,600	33%
South Carolina	14,152	17,100	83%
South Dakota	21,511	27,189	79%
Tennessee	2,181	5,600	39%
Texas	111,360	216,550	51%
Utah	3,455	6,722	51%
Vermont	1,251	3,900	32%
Virginia	17,294	30,251	57%
Washington	519,303	760,630	68%
West Virginia	2,538	5,950	43%
Wisconsin	170,503	316,800	54%
Wyoming	26,346	32,350	81%
United States	*7,252,820*	*9,661,557*	*75%*

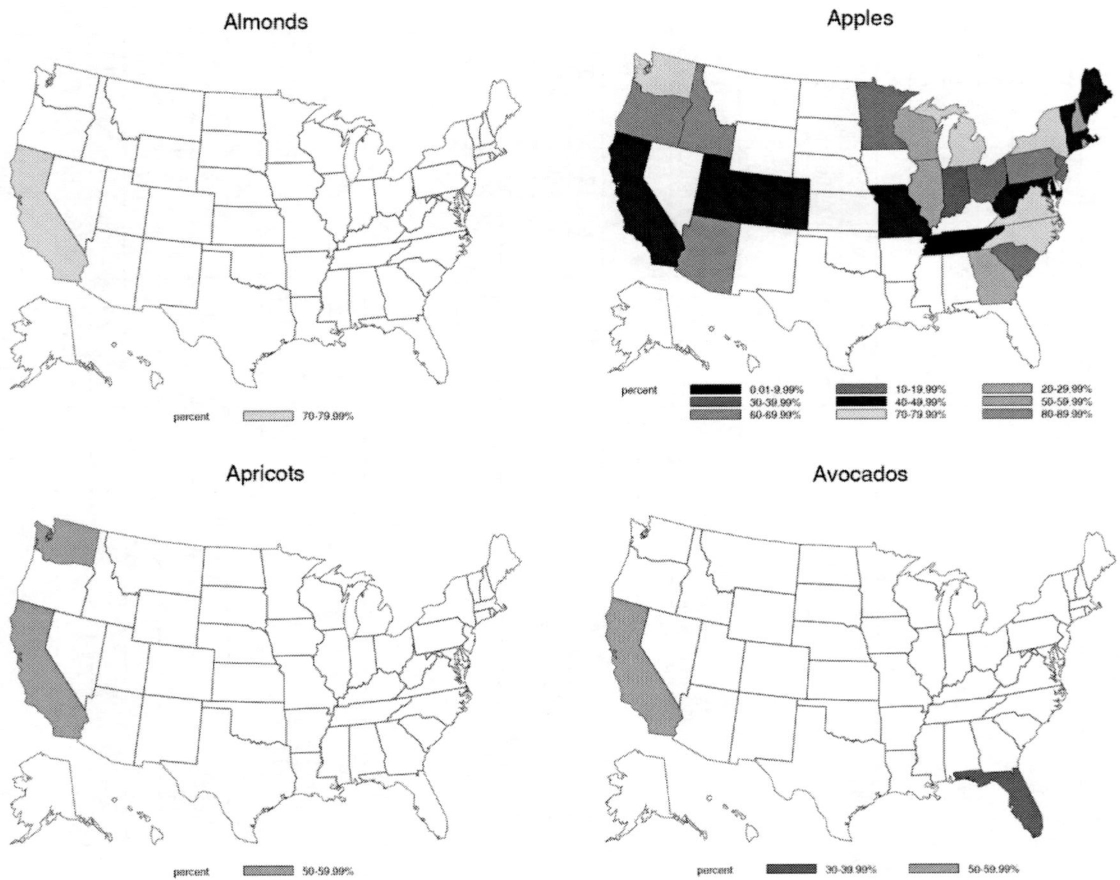
Almonds
percent
70-79.99%
Apples
percent
0.01-9.99%
30-39.99%
60-69.99%
10-19.99%
40-49.99%
70-79.99%
20-29.99%
50-59.99%
80-89.99%
Apricots
percent
50-59.99%
Avocados
percent
30-39.99%
50-59.99%

Blueberries

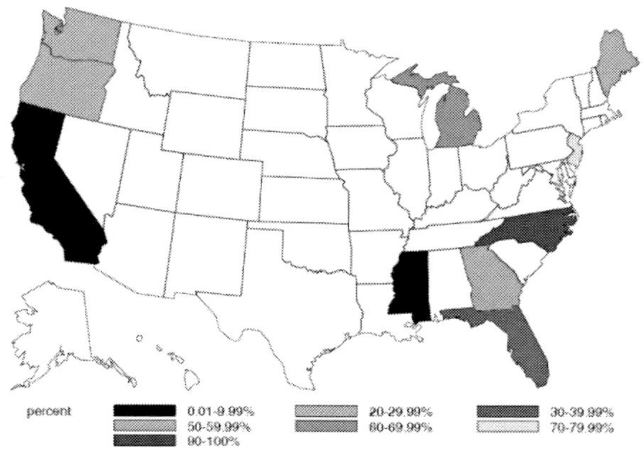

Cabbage

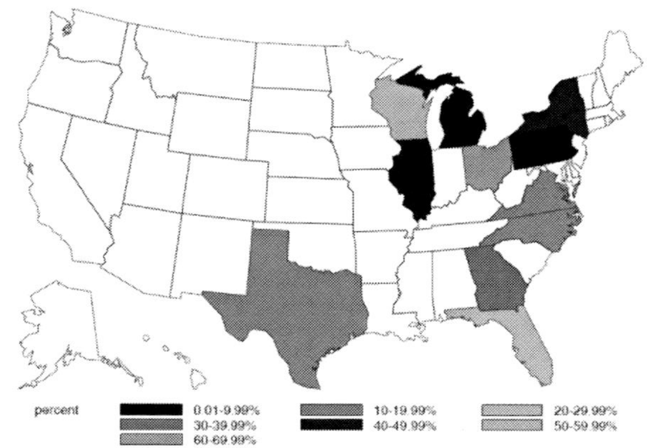

Cherries

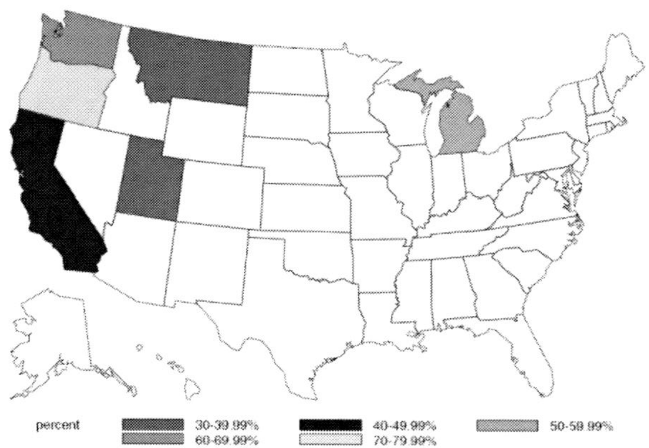

Chile Peppers

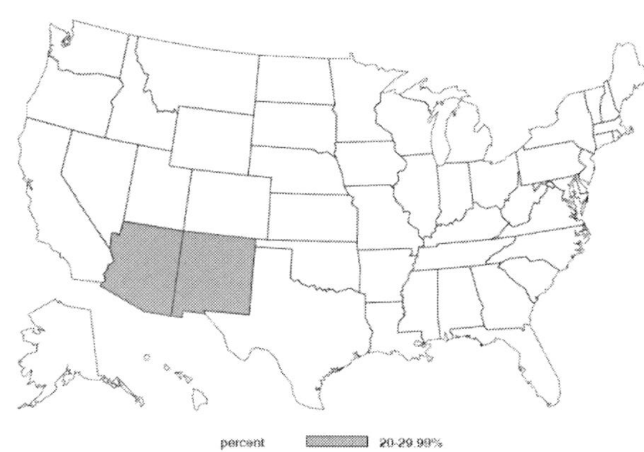

Citrus Fruit

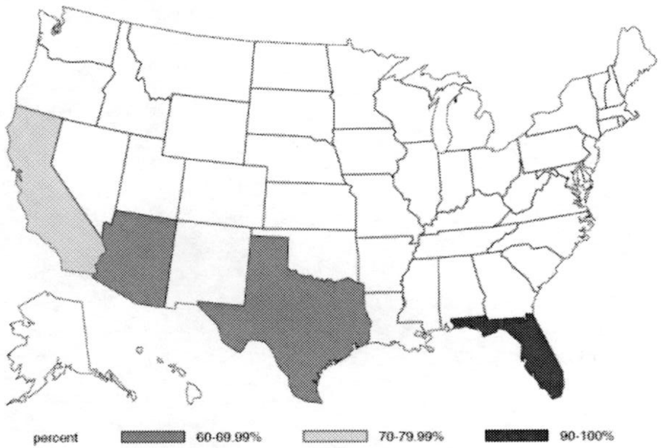

Citrus Trees

Cranberries

Dry Beans

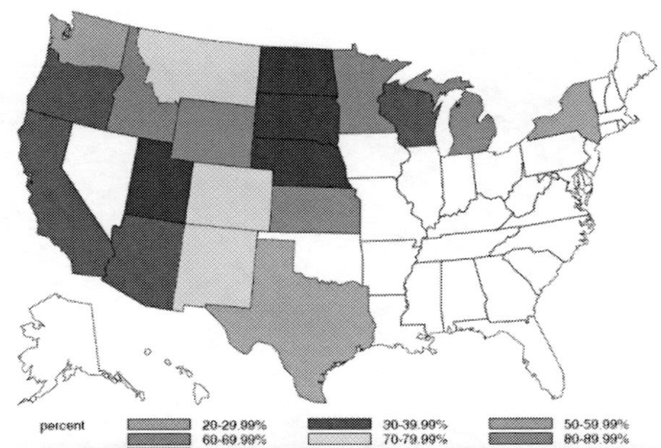

Dry Peas

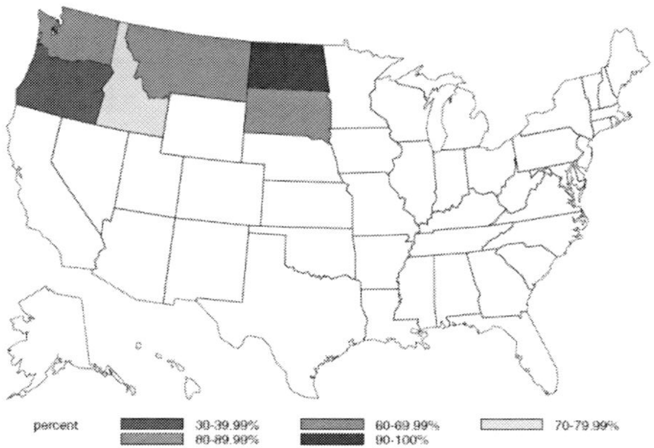

Figs

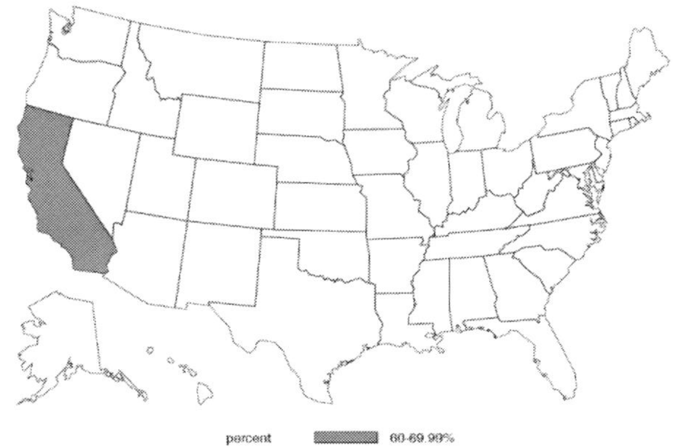

Grapes

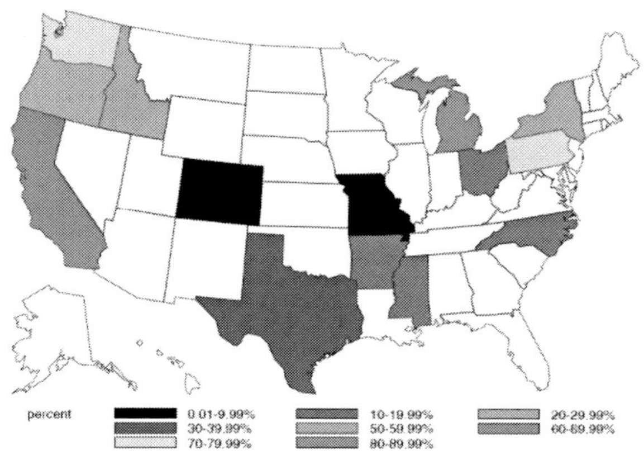

Green Peas

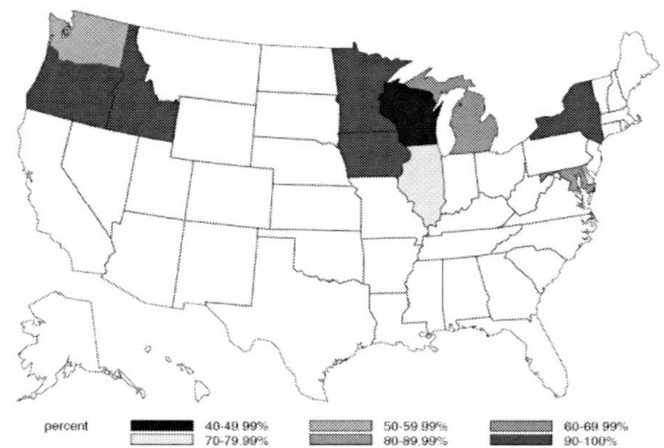

Macadamia Nuts

Nectarines

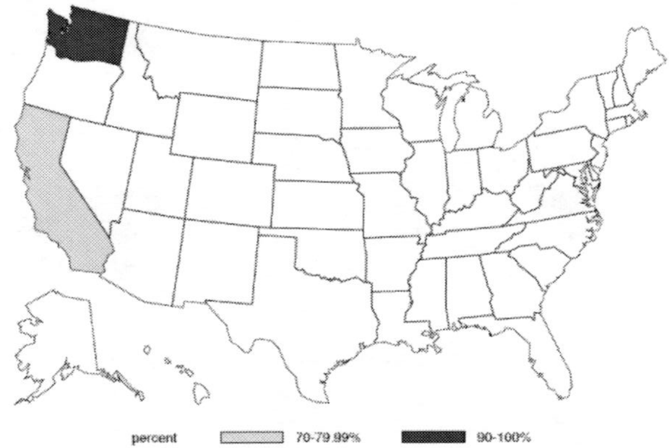

Onions

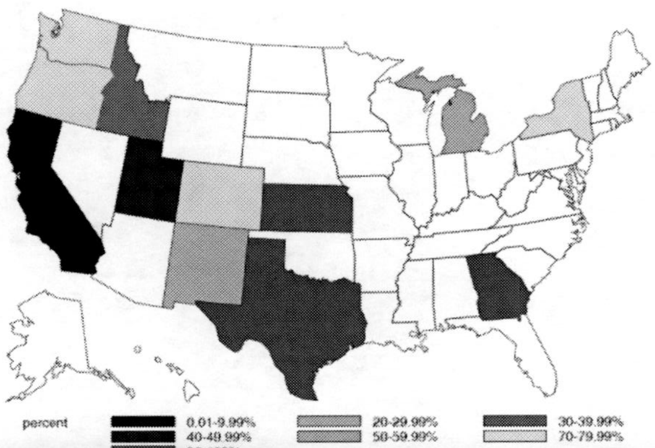

Peaches

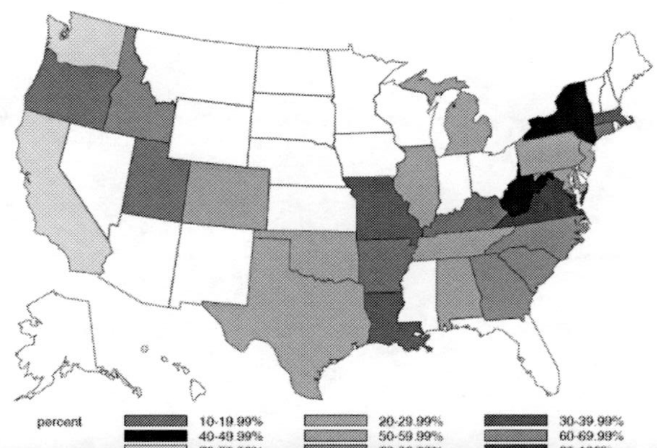

Pears

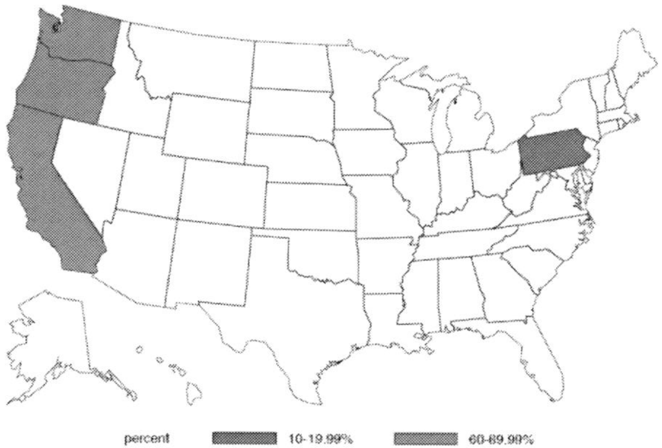

Pecans

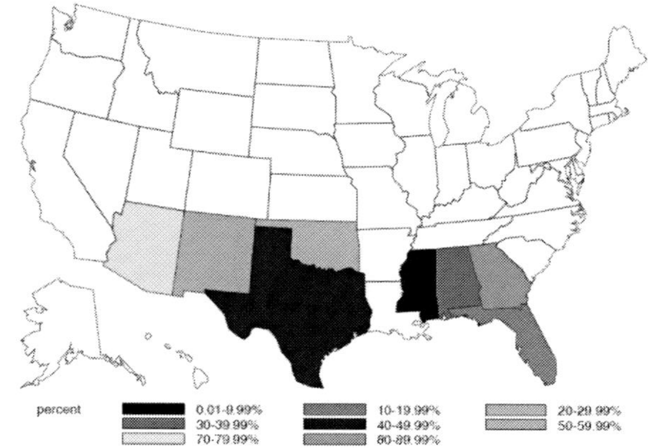

Peppers

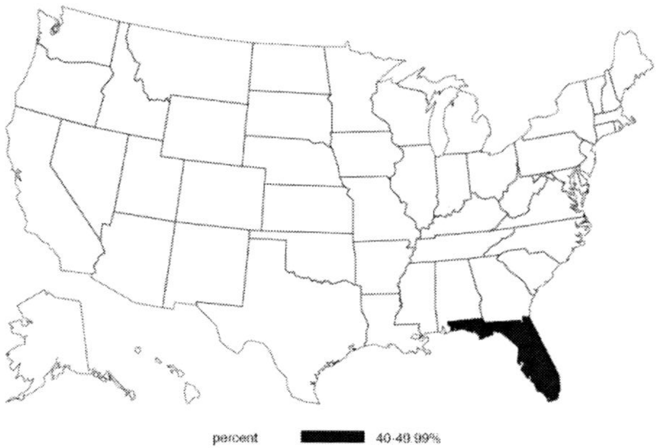

Plums

Potatoes

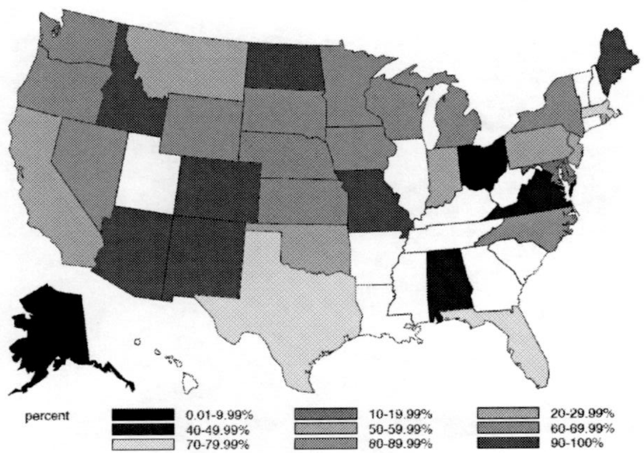

Processing Beans

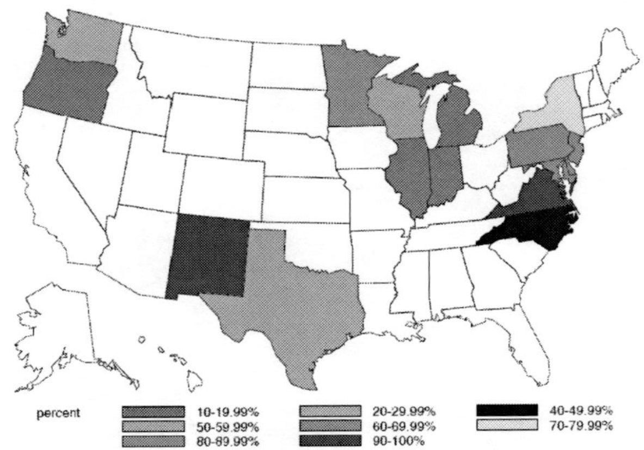

Prunes

Pumpkins

Raisins

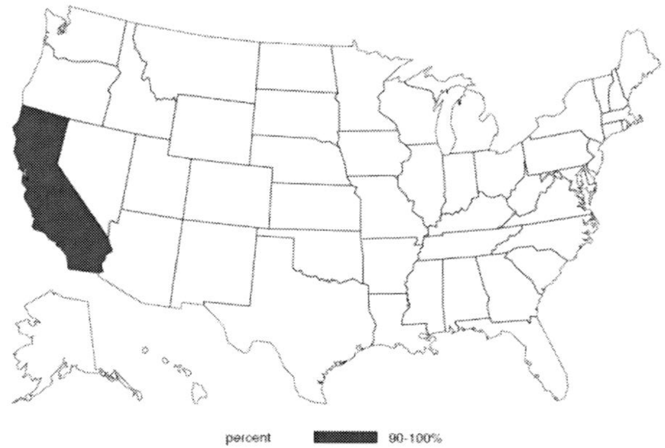

Sweet Corn

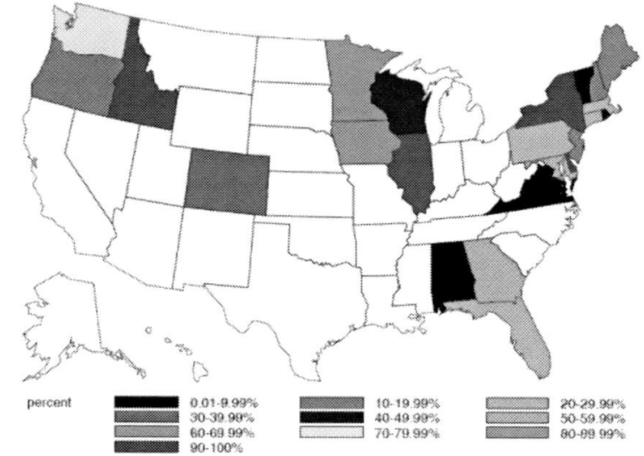

Table Grapes

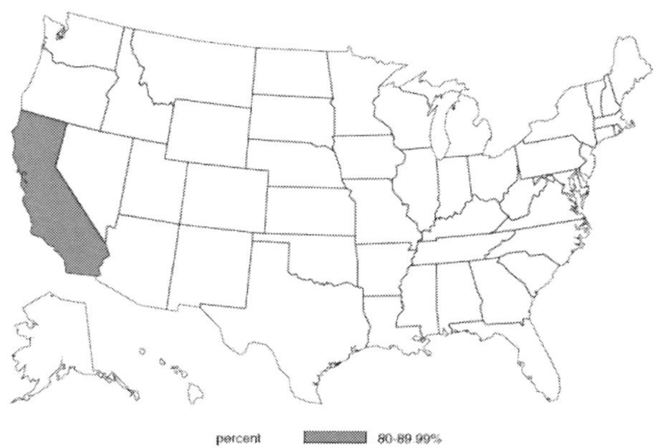

Tomatoes

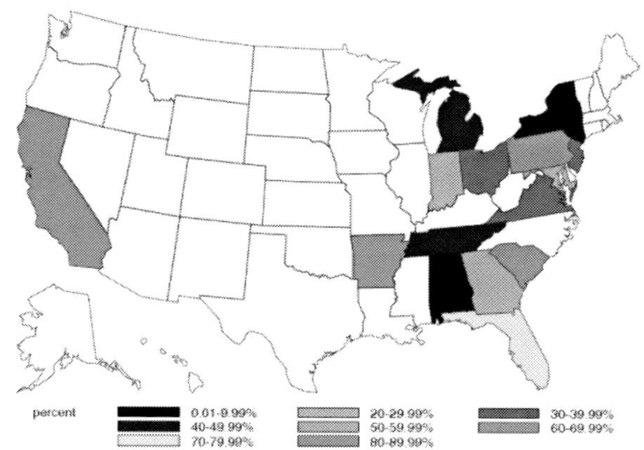

Tropical Fruit

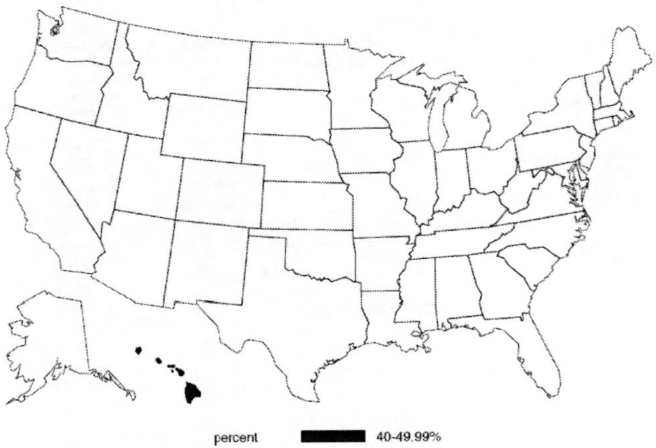

Walnuts

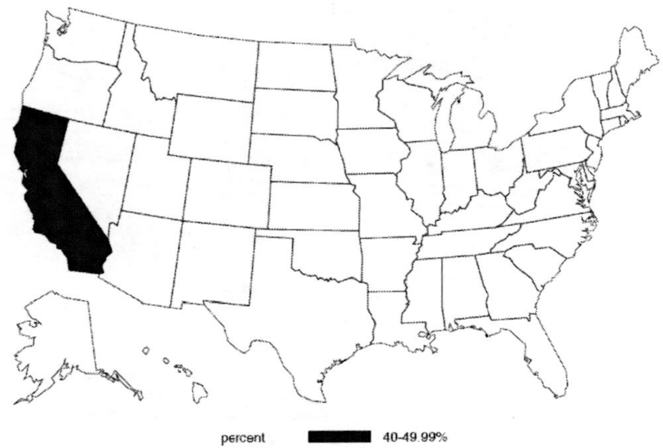

	Selected Crops in Acres																				
	Almonds			Apples			Apricots			Avocados			Blueberries			Cabbage			Cherries		
	RMA	NASS	Pct	RMA	NASS	Pct	RMA	NASS	Pct	RMA	NASS	Pct	RMA	NASS	Pct	RMA	NASS	Pct	RMA	NASS	Pct
Alabama														320							
Alaska																					
Arizona				809	1,200	67%											2,800				
Arkansas																					
California	478,380	680,000	70%	8,324	19,500	43%	6,406	11,100	58%	34,887	65,000	54%	203	2,500	8%		13,900		13,122	27,000	49%
Colorado				562	1,400	40%											2,400				
Connecticut				885	2,200	40%															
Delaware																					
Florida										2,894	7,500	39%	1,130	3,000	38%	2,354	10,400	23%			
Georgia				163	600	27%							4,830	9,500	51%	2,035	6,600	31%			
Hawaii																					
Idaho				2,198	2,600	85%														1,000	
Illinois				592	2,500	24%										0	400	0%			
Indiana				693	2,100	33%															
Iowa																					
Kansas																					
Kentucky																					
Louisiana																					
Maine				1,534	3,100	49%							13,741	23,000	60%						
Maryland				920	1,900	48%															
Massachusetts				1,793	4,000	45%															
Michigan				27,756	36,500	76%							11,337	18,600	61%	156	2,500	6%	4,064	7,200	56%
Minnesota				408	3,000	14%															
Mississippi													230	2,700	9%						
Missouri				872	1,900	46%															
Montana																			227	710	32%
Nebraska																					
Nevada																					
New Hampshire				1,158	2,100	55%															

(Continued)

(Continued)

	Selected Crops in Acres																				
	Chile Peppers			Citrus Fruit			Citrus Trees			Cranberries			Dry Beans			Dry Peas			Figs		
	RMA	NASS	Pct	RMA	NASS	Pct	RMA	NASS	Pct	RMA	NASS	Pct	RMA	NASS	Pct	RMA	NASS	Pct	RMA	NASS	Pct
Alabama																					
Alaska																					
Arizona	879	3,500	25%	10,923	17,300	63%							2,066	5,800	36%						
Arkansas																					
California				214,414	270,000	79%							20,453	52,000	39%				6,002	9,400	64%
Colorado													35,489	48,000	74%						
Connecticut																					
Delaware																					
Florida				542,441	542,441	100%															
Georgia																					
Hawaii																					
Idaho													48,078	80,000	60%	53,067	75,000	71%			
Illinois																					
Indiana																					
Iowa																					
Kansas													4,054	6,000	68%						
Kentucky																					
Louisiana																					
Maine																					
Maryland																					
Massachusetts										10,842	13,500	80%									
Michigan													168,275	200,000	84%						
Minnesota													123,547	150,000	82%						
Mississippi																					
Missouri																					
Montana													7,885	11,200	70%	280,299	328,000	85%			
Nebraska													123,890	135,000	92%						
Nevada																					
New Hampshire																					

	Selected Crops in Acres																				
	Grapes			Green Peas			Macadamia Nuts			Nectarines			Onions			Peaches			Pears		
	RMA	NASS	Pct	RMA	NASS	Pct	RMA	NASS	Pct	RMA	NASS	Pct	RMA	NASS	Pct	RMA	NASS	Pct	RMA	NASS	Pct
Alabama																1,465	2,500	59%			
Alaska																					
Arizona														1,500							
Arkansas	123	700	18%													232	1,300	18%			
California	474,194	689,000	69%							21,711	31,000	70%	3,008	47,000	6%	43,039	56,000	77%	10,288	15,000	69%
Colorado	109	1,170	9%										6,104	8,000	76%	1,269	2,300	55%			
Connecticut																50	400	13%			
Delaware				3,935	5,600	70%															
Florida																					
Georgia													11,305	12,000	94%	8,111	9,500	85%			
Hawaii							11,837	15,000	79%												
Idaho	827	1,490	56%	6,136	6,136	100%							3,413	8,800	39%	765	1,200	64%			
Illinois				7,309	9,500	77%										841	1,600	53%			
Indiana																					
Iowa				1,296	1,350	96%															
Kansas													321	321	100%						
Kentucky																56	380	15%			
Louisiana																76	240	32%			
Maine																					
Maryland				2,388	2,925	82%										315	580	54%			
Massachusetts																139	430	32%			
Michigan	11,622	14,200	82%	1,604	2,650	61%							2,289	4,000	57%	2,439	4,300	57%			
Minnesota				68,902	74,800	92%															
Mississippi	124	650	19%																		
Missouri	91	1,500	6%													637	1,700	37%			
Montana																					
Nebraska																					
Nevada														2,600							
New Hampshire																					

(Continued)

(Continued)

	Selected Crops in Acres																				
	Pecans			Peppers			Plums			Potatoes			Processing Beans			Prunes			Pumpkins		
	RMA	NASS	Pct	RMA	NASS	Pct	RMA	NASS	Pct	RMA	NASS	Pct	RMA	NASS	Pct	RMA	NASS	Pct	RMA	NASS	Pct
Alabama	4,118	11,000	37%							623	1,300	48%									
Alaska										18	800	2%									
Arizona	9,093	12,350	74%							3,317	3,500	95%									
Arkansas																					
California							21,795	29,500	74%	22,898	38,400	60%				61,752	64,000	96%			
Colorado										56,113	61,600	91%									
Connecticut																					
Delaware										1,143	1,700	67%	11,273	12,700	89%						
Florida	602	5,500	11%	9,204	19,000	48%				20,403	28,500	72%									
Georgia	82,542	96,250	86%																		
Hawaii																					
Idaho										276,353	305,000	91%									
Illinois											5,500		2,410	12,700	19%				6,188	13,100	47%
Indiana										549	2,300	24%	790	5,300	15%						
Iowa										577	650	89%									
Kansas										3,193	5,000	64%									
Kentucky																					
Louisiana																					
Maine										53,825	56,000	96%									
Maryland										481	2,500	19%	3,136	5,100	61%						
Massachusetts										2,149	2,800	77%									
Michigan										28,780	43,000	67%	2,936	15,500	19%						
Minnesota										40,861	50,000	82%	3,090	4,900	63%						
Mississippi	208	10,500	2%																		
Missouri										6,776	7,200	94%									
Montana										3,048	10,900	28%									
Nebraska										13,612	19,500	70%									
Nevada										5,075	5,800	88%									
New Hampshire																					

	Selected Crops in Acres																		All		
	Raisins			Sweet Corn			Table Grapes			Tomatoes			Tropical Fruit			Walnuts					
	RMA	NASS	Pct	RMA	NASS	Pct	RMA	NASS	Pct	RMA	NASS	Pct	RMA	NASS	Pct	RMA	NASS	Pct	RMA	NASS	Pct
Alabama				0	1,900	0%				88	1,300	7%							6,294	18,320	34%
Alaska																			18	800	2%
Arizona																			27,087	47,950	56%
Arkansas										865	1,000	87%							1,220	3,000	41%
California	249,160	249,160	100%		25,000		84,691	97,000	87%	278,975	319,000	87%				110,071	223,000	49%	2,163,773	3,033,460	71%
Colorado				2,311	7,400	31%													101,957	132,270	77%
Connecticut				931	4,500	21%													1,866	7,100	26%
Delaware				3,812	3,812	100%													20,163	23,812	85%
Florida				24,612	45,300	54%				25,094	32,400	77%							628,734	694,041	91%
Georgia				14,984	25,000	60%				2,328	4,300	54%							126,298	163,750	77%
Hawaii													3,784	8,780	43%				15,621	23,780	66%
Idaho				12,379	12,379	100%													403,216	493,605	82%
Illinois				8,997	8,997	100%													26,337	54,297	49%
Indiana					5,800					5,092	10,000	51%							7,124	25,500	28%
Iowa				2,305	3,550	65%													4,178	5,550	75%
Kansas																			7,568	11,321	67%
Kentucky																			56	380	15%
Louisiana																			76	240	32%
Maine				326	2,000	16%													69,426	84,100	83%
Maryland				2,608	11,100	23%				154	750	21%							10,002	24,855	40%
Massachusetts				1,216	5,400	23%													16,139	26,130	62%
Michigan					9,000					2,491	5,600	44%							263,749	363,050	73%
Minnesota				107,376	127,900	84%													344,184	410,600	84%
Mississippi																			562	13,850	4%
Missouri																			8,376	12,300	68%
Montana																			291,459	350,810	83%
Nebraska																			137,502	154,500	89%
Nevada																			5,075	8,400	60%
New Hampshire				209	1,700	12%													1,367	3,800	36%

(Continued)

	Selected Crops in Acres																				
	Almonds			Apples			Apricots			Avocados			Blueberries			Cabbage			Cherries		
	RMA	NASS	Pct	RMA	NASS	Pct	RMA	NASS	Pct	RMA	NASS	Pct	RMA	NASS	Pct	RMA	NASS	Pct	RMA	NASS	Pct
New Jersey				392	2,000	20%							5,400	7,600	71%		1,600				
New Mexico																					
New York				30,667	42,000	73%										4,231	10,100	42%			
North Carolina				4,966	6,800	73%							5,016	5,016	100%	883	5,300	17%			
North Dakota																					
Ohio				1,319	7,000	19%										721	1,200	60%			
Oklahoma																					
Oregon				2,627	4,200	63%							1,134	5,200	22%				8,803	12,500	70%
Pennsylvania				14,193	21,000	68%										69	1,200	6%			
Rhode Island				86	300	29%															
South Carolina				70	500	14%															
South Dakota																					
Tennessee				51	900	6%															
Texas																1,333	8,200	16%			
Utah				629	1,400	45%													162	500	32%
Vermont				1,156	2,800	41%															
Virginia				8,768	12,000	73%										62	500	12%			
Washington				120,149	153,000	79%	735	1,280	57%				1,034	4,100	25%				23,006	33,000	70%
West Virginia				2,075	5,000	42%															
Wisconsin				941	4,700	20%										1,837	3,100	59%			
Wyoming																					
United States	478,380	680,000	70%	236,756	348,200	68%	7,141	12,380	58%	37,781	72,500	52%	44,055	81,536	54%	13,681	70,200	19%	49,384	81,910	60%

	Selected Crops in Acres																				
	Chile Peppers			Citrus Fruit			Citrus Trees			Cranberries			Dry Beans			Dry Peas			Figs		
	RMA	NASS	Pct	RMA	NASS	Pct	RMA	NASS	Pct	RMA	NASS	Pct	RMA	NASS	Pct	RMA	NASS	Pct	RMA	NASS	Pct
New Jersey										2,897	3,100	93%									
New Mexico	2,892	12,300	24%										7,428	9,300	80%						
New York													8,652	17,000	51%						
North Carolina																					
North Dakota													650,859	660,000	99%	569,973	615,000	93%			
Ohio																					
Oklahoma																					
Oregon										1,096	2,700	41%	1,716	4,800	36%	2,132	5,500	39%			
Pennsylvania																					
Rhode Island																					
South Carolina																					
South Dakota													8,839	8,839	100%	11,973	17,500	68%			
Tennessee																					
Texas				18,508	27,300	68%	21,144	27,300	77%				6,594	24,000	27%						
Utah													1,572	1,572	100%						
Vermont																					
Virginia																					
Washington										684	1,700	40%	29,053	50,000	58%	81,688	130,000	63%			
West Virginia																					
Wisconsin										16,097	17,500	92%	2,171	6,500	33%						
Wyoming													25,587	31,500	81%						
United States	3,771	15,800	24%	786,286	857,041	92%	21,144	27,300	77%	31,616	38,500	82%	1,276,208	1,501,511	85%	999,132	1,171,000	85%	6,002	9,400	64%

(Continued)

	Selected Crops in Acres																				
	Grapes			Green Peas			Macadamia Nuts			Nectarines			Onions			Peaches			Pears		
	RMA	NASS	Pct	RMA	NASS	Pct	RMA	NASS	Pct	RMA	NASS	Pct	RMA	NASS	Pct	RMA	NASS	Pct	RMA	NASS	Pct
New Jersey																3,704	6,200	60%			
New Mexico													1,488	7,000	21%						
New York	19,435	37,000	53%	20,119	20,119	100%							7,985	10,600	75%	669	1,600	42%			
North Carolina	248	1,700	15%													721	1,200	60%			
North Dakota																					
Ohio	308	1,900	16%																		
Oklahoma																237	800	30%			
Oregon	3,876	14,900	26%	18,647	18,700	100%							15,585	21,800	71%	127	650	20%	10,139	16,200	63%
Pennsylvania	9,875	13,600	73%													2,369	4,400	54%	112	800	14%
Rhode Island																					
South Carolina																11,933	14,000	85%			
South Dakota																					
Tennessee																262	500	52%			
Texas	873	2,500	35%										11,725	11,800	99%	1,352	4,900	28%			
Utah													849	1,750	49%	243	1,500	16%			
Vermont																					
Virginia																1,179	1,179	100%			
Washington	45,787	58,000	79%	22,164	41,900	53%				1,298	1,300	100%	15,968	21,900	73%	1,661	2,300	72%	14,565	24,000	61%
West Virginia																463	950	49%			
Wisconsin				20,286	40,700	50%								1,900							
Wyoming																					
United States	567,492	838,310	68%	172,786	224,380	77%	11,837	15,000	79%	23,009	32,300	71%	80,040	160,971	50%	84,354	122,609	69%	35,104	56,000	63%

	Selected Crops in Acres																				
	Pecans			Peppers			Plums			Potatoes			Processing Beans			Prunes			Pumpkins		
	RMA	NASS	Pct	RMA	NASS	Pct	RMA	NASS	Pct	RMA	NASS	Pct	RMA	NASS	Pct	RMA	NASS	Pct	RMA	NASS	Pct
New Jersey										417	2,000	21%	662	1,050	63%						
New Mexico	20,880	39,250	53%							5,699	5,900	97%	852	852	100%						
New York										11,167	18,000	62%	16,835	21,500	78%						
North Carolina										12,092	14,500	83%	1,327	3,050	44%						
North Dakota										76,680	82,000	94%									
Ohio										35	2,500	1%									
Oklahoma	4,932	21,600	23%							156	300	52%									
Oregon										23,766	35,300	67%	2,191	18,700	12%						
Pennsylvania										5,003	10,000	50%	7,162	10,700	67%						
Rhode Island										108	500	22%									
South Carolina																					
South Dakota										699	850	82%									
Tennessee																					
Texas	34,598	83,000	42%							11,594	16,400	71%	3,639	7,250	50%						
Utah																					
Vermont																					
Virginia										2,833	5,800	49%	2,672	2,672	100%						
Washington										101,380	155,000	65%	1,233	5,050	24%						
West Virginia																					
Wisconsin										41,880	63,500	66%	42,334	82,300	51%						
Wyoming										759	850	89%									
United States	156,973	279,450	56%	9,204	19,000	48%	21,795	29,500	74%	834,062	1,065,350	78%	102,542	209,324	49%	61,752	64,000	96%	6,188	13,100	47%

(Continued)

	Selected Crops in Acres																		All		
	Raisins			Sweet Corn			Table Grapes			Tomatoes			Tropical Fruit			Walnuts					
	RMA	NASS	Pct	RMA	NASS	Pct	RMA	NASS	Pct	RMA	NASS	Pct	RMA	NASS	Pct	RMA	NASS	Pct	RMA	NASS	Pct
New Jersey				996	7,100	14%				918	3,000	31%							15,386	33,650	46%
New Mexico																			39,239	74,602	53%
New York				25,257	25,257	100%				18	3,000	1%							145,035	206,176	70%
North Carolina					6,800						3,500								25,253	47,866	53%
North Dakota																			1,297,512	1,357,000	96%
Ohio					16,300					3,912	10,800	36%							6,295	39,700	16%
Oklahoma																			5,325	22,700	23%
Oregon				3,675	23,200	16%													95,514	184,350	52%
Pennsylvania				3,475	15,700	22%				1,162	1,900	61%							43,420	79,300	55%
Rhode Island				326	800	41%													520	1,600	33%
South Carolina										2,149	2,600	83%							14,152	17,100	83%
South Dakota																			21,511	27,189	79%
Tennessee										1,868	4,200	44%							2,181	5,600	39%
Texas					2,800						1,100								111,360	216,550	51%
Utah																			3,455	6,722	51%
Vermont				95	1,100	9%													1,251	3,900	32%
Virginia				1	3,300	0%				1,779	4,800	37%							17,294	30,251	57%
Washington				58,898	78,100	75%													519,303	760,630	68%
West Virginia																			2,538	5,950	43%
Wisconsin				44,957	96,600	47%													170,503	316,800	54%
Wyoming																			26,346	32,350	81%
United States	249,160	249,160	100%	319,746	577,795	55%	84,691	97,000	87%	326,893	409,250	80%	3,784	8,780	43%	110,071	223,000	49%	7,252,820	9,661,557	75%

INDEX

A

access, 40
accounting, 17
Actual Revenue History (ARH), xi, 76, 94, 99
AD, 31
additives, 23
adjustment, 15, 31, 83, 100
administrative support, 31
adverse weather, 15, 91, 108
agencies, viii, 2, 10, 11, 24, 26, 27, 31, 34, 37, 46, 48, 50, 51, 54, 61, 64, 95, 103
agricultural exports, 3, 28, 29, 33, 36
agricultural producers, viii, xi, 2, 8, 10, 13, 15, 37, 38, 40, 48, 52, 53, 56, 94, 106
Agricultural Research Service, 10, 37
Agricultural Research Service (ARS), 10, 37
agriculture, 3, 8, 11, 17, 19, 22, 26, 28, 32, 33, 34, 36, 38, 40, 46, 48, 50, 51, 59, 64, 91
airports, 35
Alaska, 114
alfalfa, 24
almonds, 12, 21, 41, 44, 90
alters, 92
animal welfare, 32
antidumping, viii, 2, 31
apples, 12, 59, 74, 90
appropriations, 8, 15, 16, 19, 20, 21, 33, 36, 37, 39, 42, 49, 51, 55, 59
Appropriations Committee, 45
aquaculture, 55, 60, 109
Argentina, 32
arrest, 33
ARS, 34, 37, 39, 62
Asian longhorned beetle, 33, 35
asparagus, 12, 16, 29, 31, 41, 44, 59, 90
assessment, 21, 98, 105
assets, 16
audit, 26, 34
authorities, 19, 32, 33, 60
authority, 14, 25, 33, 63, 87, 88, 96, 97, 98, 99, 107
average revenue, 77, 87
awareness, 19

B

background information, 57
bacteria, 27
baggage, 35
bankruptcy, 16
banks, 29, 71
barriers, 30, 37, 60

base, 17, 52, 56, 77, 89
benefits, vii, 1, 8, 20, 31, 46, 47, 52, 88, 102
beverages, 23, 57
biofuel, 104
biological control, 32
biomass, 87
bonuses, 41
border crossing, 35
border inspections, 26
Brazil, 32
breeding, 39, 51
bulk commodity crops, vii, 1, 3, 30
Bureau of Customs and Border Protection, 36
business strategy, 47
businesses, 18
buyer, 16
buyers, 16, 29, 83

C

cabbage, 12, 21, 26, 59, 90
capacity building, 32
cash, 19, 31, 43, 45, 46, 56, 63, 83, 91, 104
Catalog of Federal Domestic Assistance, 11
category a, 45
CBP, 25, 26, 35, 36
CDC, 24, 27
Census, 4, 5, 6, 9, 38, 54, 55, 61, 91
certification, 17, 21, 34, 36, 47, 49, 50, 52, 53, 64, 81
CFR, 56, 57, 58, 59, 60, 61, 64, 66
challenges, 14, 76, 108, 109
chemical, 28, 59
Chicago, 99
children, 27, 28, 40, 41, 43, 45, 46, 54
Chile, 32, 79, 107, 113
China, 32
classification, 95
Clean Air Act, 39, 62
coffee, 90
collaboration, 33
colleges, 38
color, 23
commerce, 16
commercial, 3, 14, 17, 19, 21, 22, 28
commercial crop, 14
commodity, vii, xi, 1, 3, 4, 6, 8, 12, 13, 14, 15, 18, 20, 22, 23, 30, 31, 40, 41, 42, 43, 48, 52, 56, 60, 62, 70, 73, 78, 80, 84, 88, 91, 94, 96, 98, 99, 103, 105
commodity futures, xi, 94, 99
commodity producers, 18
communication, 65
community, 19, 35, 41
competing interests, xi, 94
competition, 31
competitive grant program, 38
competitiveness, 17, 51
compilation, 38, 91
complexity, 14, 83
compliance, 21, 63, 84, 106, 107
computer, 82
Congress, v, ix, x, 1, 6, 8, 13, 16, 25, 27, 34, 45, 50, 55, 59, 61, 65, 67, 68, 69, 70, 76, 78, 79, 83, 84, 86, 91, 93
conservation, 8, 10, 52, 56
consumers, 19, 21, 22, 23, 24, 27, 47, 49
consumption, viii, 2, 3, 24, 28, 41, 47, 59, 62
consumption rates, 28
contamination, x, 24, 26, 27, 68, 85, 87
contingency, 41
control measures, 90
cooperation, 26, 28
cooperative agreements, 22, 34
Cooperative State Research, Education, and Extension Service, 61
cost, x, 14, 28, 30, 32, 33, 43, 46, 50, 53, 64, 74, 76, 83, 87, 89, 93, 97, 107
cotton, vii, 1, 14, 51, 55, 57, 70, 82, 95
country of origin, 23
covering, 18, 33, 36, 38

creditors, 16
crop insurance, vii, ix, x, xi, 1, 8, 12, 13, 14, 15, 55, 67, 68, 69, 70, 72, 73, 75, 77, 78, 80, 81, 83, 85, 86, 87, 88, 90, 93, 94, 95, 96, 97, 101, 102, 103, 104, 106, 107, 108
crop producers, vii, ix, 1, 10, 15, 51, 52, 67, 68, 69, 85, 86
crop production, 3, 13, 83, 99
crop rotations, 83
CRP, 53
CSA, 19
CSREES, 61
cultivars, 72, 103
customers, 23
Customs and Border Protection, 25
CVD, 31
cyst, 35

D

data collection, 20, 38, 103
database, 24, 28, 31, 35, 36, 54, 58, 61
Department of Agriculture, v, vii, 1, 3, 60, 93
Department of Commerce, viii, 2, 11, 31, 32
Department of Defense, 42
Department of Education, 54
Department of Health and Human Services, viii, 2, 24, 62
Department of Homeland Security, viii, 2, 36, 53
Department of Labor, viii, 2, 53
destruction, 16, 33, 73, 78
detection, 32, 34, 35, 61
developing countries, 30
DHS, 36, 53, 61
directors, 69, 72
disaster, vii, 1, 8, 15, 16, 42, 46, 69, 72, 85, 89, 107
disaster area, 16
disaster assistance, vii, 1, 8, 15, 72, 85, 107
discretionary funding, viii, 2
diseases, 26, 32, 33, 34, 35, 36, 39
distribution, 3, 22, 27, 40, 41, 53, 59, 63
District of Columbia, 17
diversification, 87
diversity, 75
DOL, 53, 54, 66
domestic industry, 31
donations, 41
draft, 58, 61
drought, ix, 68, 73
dumping, viii, 2, 32

E

early warning, 37
economic damage, 35
economic development, 19
economic losses, 99
economic problem, 41
Economic Research Service, 37, 51, 91
Economic Research Service (ERS), 37, 51
economics, 51
education, 18, 34, 38, 46, 51, 54, 62, 64, 101
elementary school, 46
eligibility criteria, 43
emerald ash borer, 35
emergency, 16, 32, 33, 35, 41
Emergency Assistance, 9, 15, 56
emergency response, 32
employees, 22
employers, 53
employment, 53, 54
energy, 87
enforcement, 17, 24, 49
environment, 27, 100, 109
Environmental Protection Agency, viii, 2, 11, 25
EPA, 25, 27, 28, 62

equipment, 16, 19, 35
ERS, 38, 66
ETA, 54, 66
ethanol, 104
EU, 32
evidence, xi, 94
exclusion, 32, 36
expenditures, 46
expertise, 22, 23, 27
export market, 28
export promotion, 29, 59
exporters, 28, 29, 36
exports, 29, 30, 36, 37, 61
exposure, 27, 72, 98
extension services, vii, 1, 8, 37, 38

F

families, 46, 54
farm bill, viii, x, 2, 4, 6, 7, 8, 15, 16, 17, 18, 19, 20, 23, 28, 29, 30, 34, 39, 42, 45, 46, 48, 50, 51, 52, 54, 56, 62, 63, 64, 65, 68, 69, 85, 86, 90, 92
Farm Bill, 9, 48, 52, 55, 61, 86, 87, 88, 89
farmers, vii, viii, 1, 2, 16, 19, 20, 21, 22, 25, 31, 38, 47, 50, 64, 67, 70, 71, 81, 83, 88
farms, viii, 3, 4, 19, 25, 47, 48, 49, 53, 58, 67, 69
FAS, 10, 28, 29, 30, 31, 41, 56, 59, 60
FDA, viii, 2, 21, 22, 23, 24, 25, 26, 27, 28, 58
federal agency, 22, 24, 25
federal assistance, 55
Federal Crop Insurance Corporation (FCIC), ix, x, 55, 68, 69, 93
federal crop insurance program, viii, ix, 11, 14, 67, 68, 69, 75, 86, 89, 90
federal funds, 38, 43
federal government, ix, 12, 68, 70, 73, 80
Federal Register, 11, 21, 55, 57, 58, 61, 65, 97
fiber, 55
field crops, 64, 75, 83
financial, ix, 15, 29, 50, 52, 65, 67, 72, 80, 83, 84, 86
financial institutions, 29
financial resources, 84
fish, 24, 31, 55, 89
fisheries, 25
flexibility, 52
flowers, 15
fluctuations, 109
food, viii, x, 2, 8, 11, 18, 20, 21, 22, 23, 24, 25, 26, 27, 30, 38, 39, 40, 41, 42, 43, 45, 46, 55, 58, 59, 62, 63, 64, 68, 85, 87
food additive, 23
food additives, 23
Food and Drug Administration, viii, 2, 11, 22, 24
food industry, 22
food safety, viii, x, 2, 11, 18, 21, 22, 24, 25, 27, 38, 39, 59, 68, 85, 87
Food, Conservation, and Energy Act of 2008, viii, 2, 54, 55, 97
foodborne illness, 24, 25
force, 60
foreign banks, 29
foreign firms, 26
formation, 39, 62
free trade, 30
fresh market beans, xi, 90, 94, 95
frost, ix, 68, 73, 89
fruits, viii, ix, 2, 3, 4, 8, 11, 12, 13, 14, 16, 20, 22, 23, 25, 26, 28, 30, 36, 40, 41, 42, 43, 45, 46, 47, 48, 51, 52, 54, 55, 57, 58, 62, 63, 64, 67, 69, 74, 81, 90, 91, 95
funding, viii, ix, xi, 2, 4, 6, 7, 8, 17, 19, 20, 29, 30, 33, 38, 40, 42, 46, 48, 50, 51, 55, 56, 57, 63, 64, 68, 69, 88, 89, 95
funds, 17, 19, 21, 22, 28, 29, 30, 33, 34, 35, 38, 39, 50, 51, 59, 62, 63, 64, 86

G

GAO, 90
GATT, 60
General Accounting Office, 90
General Agreement on Tariffs and Trade, 60
genetic engineering, 64
genetics, 39
genomics, 39
Georgia, 56, 101, 114
Germany, 32
ginseng, 15, 23, 29
global trade, 11
governance, 21
governments, 33, 95
grading, 17, 21, 22, 57
grant programs, 38
grants, 17, 18, 19, 20, 38, 39, 46, 51, 53, 54, 64
grasses, 89
grazing, 89
Great Depression, 70
greenhouse, 4, 91
growth, x, 24, 28, 63, 71, 94, 102, 110
guidance, 58
guidelines, 45, 63

H

harvesting, 25, 53
Hawaii, 29, 36, 61, 114
hazards, 24, 39, 59
health, 18, 24, 27, 32, 33, 60, 61, 64
health problems, 32
HHS, 24, 62
higher education, 51
history, 64, 73, 74, 76, 81, 90
homes, 41
honesty, 23
horticultural crops, vii, 1, 3
host, 37
House, x, 13, 14, 40, 45, 55, 61, 68, 69, 85, 86, 87, 88, 89
House of Representatives, 13, 14, 40, 55
household income, 45
housing, 54, 72
human, viii, 2, 27, 28
human health, 27, 28
hygiene, 24, 25

I

identification, 34
identity, 23
immigration, 66
imported products, 23, 25, 32, 36
imports, 16, 25, 28, 31, 32, 33
improvements, 72, 109, 110
income, vii, 1, 3, 4, 6, 12, 13, 15, 16, 20, 40, 41, 47, 54, 56, 70, 75, 78, 88, 109
income support, vii, 1, 4, 6, 12, 13, 56, 78, 88
India, 32
Indian reservation, 46
individuals, 17, 40, 42, 46
Indonesia, 32
industries, x, 22, 32, 94, 104, 106
industry, vii, viii, xi, 1, 2, 6, 8, 11, 16, 17, 21, 24, 29, 30, 31, 34, 35, 39, 40, 53, 58, 70, 72, 75, 82, 94, 104, 108
infants, 28
infrastructure, 19, 35
ingredients, 23
injury, 31, 32
insects, ix, 68, 73
inspections, 33, 35
inspectors, 61
institutions, 37, 54
integration, 51
integrity, 49
Internal Revenue Service, 77, 87
international standards, 58
Iowa, 52, 65, 115
irradiation, 64

islands, 36
issues, 11, 14, 26, 30, 36, 37, 38, 39, 54, 58, 82, 83, 91, 97

J

job training, 54
jurisdiction, viii, 2, 40
justification, 34, 57

L

labeling, 23, 25, 27, 58, 84
labor shortage, 53
laws, viii, 2, 8, 16, 38, 40, 59, 60
lead, 22, 38
leadership, 110
legislation, viii, 2, 8, 25, 45, 69, 95
legislative proposals, vii, ix, 68, 69
light, 33, 35
Listeria monocytogenes, 24
livestock, 3, 7, 45, 51, 56, 57, 60, 62, 86, 89
loans, 16, 34, 71
local government, 19, 34, 54
Louisiana, 61, 102, 115

M

majority, 3, 18, 21, 71, 104
management, 32, 34, 51, 59, 99
manufacturing, 59
manure, 24
market access, 19
market penetration, 107
marketability, 14, 82
marketing, vii, 1, 8, 16, 17, 18, 19, 21, 22, 29, 30, 51, 75, 99
marketing strategy, 18
marketplace, viii, 2, 8, 98
Maryland, 115
materials, 36, 49, 96, 97
meat, 8, 23, 25, 26, 48, 51, 57, 64
methodology, 99, 102, 110
Mexico, 32, 90, 102, 115
Miami, 102
military, 22, 63
mission, 17, 49
Missouri, 52, 61, 102, 115
momentum, 109
Montana, 13, 78, 102, 110, 115
Montreal Protocol, 39
moral hazard, 92
mortality, 15

N

NAP, x, 14, 15, 55, 68, 72, 85, 88, 89, 90, 107
national product, 56
National School Lunch Program, 42, 43, 62, 63
natural disaster, 15, 89
natural disasters, 15
natural resources, 38
Natural Resources Conservation Service, 10, 52
Natural Resources Conservation Service (NRCS), 10, 52
needy, 40
nematode, 35
niche market, 19
No Child Left Behind, 54, 66
NOAA, 102
Noninsured Crop Disaster Assistance Program, 55
North America, 30
North American Free Trade Agreement, 30
NRCS, 65
nursery, vii, 1, 2, 3, 4, 13, 15, 30, 34, 38, 40, 55, 61, 72, 74, 75, 79, 82, 89, 90, 91, 95, 108, 109
nutrition, vii, 1, 4, 8, 20, 38, 40, 41, 42, 43, 45, 46, 47, 62, 64

O

Office of Management and Budget, xi, 94
Office of the United States Trade Representative, 30
officials, 37
OFPA, 47, 49, 64
oil, 21
Oklahoma, 57, 102, 115
OMB, 25
operating costs, ix, 68, 70
operations, 3, 4, 21, 25, 38, 47, 48, 49, 50, 87
opportunities, 99
organic food, 29
Organic Foods Production Act, 47
organize, 35
outreach, 19, 34
outreach programs, 34
overlap, 89
oversight, 24, 41, 96
ozone, 39, 62

P

P.L. 105-220, 66
participants, 41, 64
pasture, 87, 91
pathogens, 27, 34, 36
pathways, 96
penalties, 17
Perishable Agricultural Commodities Act of 1930 (PACA), viii, 2, 8, 16
permit, 78
pest regulation, 33
pesticide, viii, 2, 27, 28, 59
pests, 26, 32, 33, 34, 35, 36, 39, 40
Philadelphia, 42
plants, 12, 22, 26, 32, 36, 54, 61, 75, 78, 91, 108
policy, 15, 55, 66, 70, 72, 73, 74, 76, 78, 80, 82, 83, 84, 85, 86, 87, 90, 96, 97, 100, 103, 105, 108, 109
pollination, 15
pollinators, 39
population, 63, 110
portfolio, 95, 96, 109
potato, 29, 113
poultry, 4, 25, 26, 45, 51, 57, 62, 64
poverty, 54
precipitation, ix, 68, 73, 102, 105
preparation, 24, 47
preparedness, 26
prevention, 32
price changes, 84
principles, 97
private sector, 71, 96, 98, 106
procurement, 42, 43, 59
product design, 108
profit, 35, 54
profitability, 39
project, 19, 39, 52, 62, 63, 65
protection, 33, 74, 77, 78, 80, 86, 90, 98, 99, 100, 106, 107, 108, 109
pruning, 89
public health, 26
public interest, 21
Puerto Rico, 22, 36, 102

Q

quality standards, 21

R

recession, 72
recommendations, 21
Reform, 14, 55, 62, 65, 66, 88, 90
reforms, viii, 2
regulations, viii, 2, 20, 22, 23, 25, 28, 36, 37, 47, 49, 55, 61, 64
regulatory agencies, 37

regulatory requirements, viii, 2
reimburse, 30, 50, 101
reinsurance, 88, 90
remedial actions, 33
renewable energy, 18
reparation, 17
reporters, 20
requirements, viii, x, xi, 2, 11, 21, 22, 23, 25, 26, 30, 36, 45, 46, 52, 54, 57, 63, 91, 93, 94, 107, 109
Reservations, 42, 46
residues, viii, 2, 27
resolution, 16, 32, 37
resources, xi, 27, 33, 47, 58, 63, 94, 107
response, x, 23, 33, 82, 93
responsiveness, 97
restaurants, 24
retail, 3, 24
revenue, viii, ix, x, xi, 12, 15, 16, 67, 68, 69, 70, 73, 75, 76, 77, 82, 84, 85, 86, 87, 94, 99, 100, 105, 108, 109
rights, 86
risk, viii, 12, 27, 28, 34, 35, 36, 67, 69, 70, 71, 72, 75, 78, 80, 82, 83, 84, 86, 87, 88, 90, 92, 98, 102, 103, 106, 107
risk assessment, 107
risk management, viii, 34, 67, 69, 75, 87, 98, 103
Risk Management Agency (RMA), ix, x, 10, 11, 68, 69, 89, 93
risk profile, 103
risks, 14, 36, 71
root, 12
rules, 25, 35, 42
rural development, 10, 38

S

safety, viii, ix, 2, 21, 22, 24, 25, 26, 27, 28, 54, 67
Salmonella, 24, 27
saturation, x, 93
savings, xi, 69, 88, 94
school, 22, 41, 42, 43, 45, 46, 63
science, 25, 26, 97, 99
SCO, 85, 86
scope, 98, 105
seafood, 23, 31, 45, 60, 62
Secretary of Agriculture, 69
seed, 31, 55
sellers, viii, 2, 8, 16
seminars, 30
Senate, x, 40, 68, 69, 85, 86, 87, 88, 89
services, vii, 1, 8, 15, 17, 37, 38, 43, 50, 53, 54, 57, 64
sewage, 64
shellfish, 24
shock, 100
shortage, 53
shortfall, 74
showing, 80, 81, 90
signs, 70
sludge, 64
SNAP, 7, 20, 46, 47, 64
social sciences, 51
solution, 110
South Africa, 32
South Dakota, 13, 78, 102, 110, 115
soybeans, ix, xi, 14, 55, 67, 82, 94, 95
Spain, 32
species, 109
spending, vii, 1, 3, 6, 8, 42, 62, 107
SS, 64
stability, 54
stakeholders, 22
standardization, 17
state, 3, 11, 17, 20, 22, 24, 26, 28, 29, 31, 33, 34, 35, 38, 41, 47, 49, 50, 54, 56, 57, 58, 59, 61, 63, 64, 65, 80, 91, 95, 98, 104, 108, 110
states, 3, 13, 17, 22, 23, 28, 29, 32, 35, 38, 41, 45, 46, 50, 52, 63, 64, 74, 75, 78, 80, 84, 85, 97, 98, 100, 108, 110
statutes, 20, 33
stimulus, 60
stock, 34, 76, 89

storage, 40
structure, 21, 82, 103
Subsidies, 70, 80
subsidized policies, ix, 67, 69
subsidy, 43, 55, 70, 80, 81, 86, 91
sugar beet, xi, 94
sugarcane, xi, 87, 94, 103
supervision, 69
Supplemental Nutrition Assistance Program, 9, 20, 46, 64
Supplemental Nutrition Assistance Program (SNAP), 9, 64
suppliers, 11, 63
surplus, 41
surveillance, 27, 33, 34

T

TAP, 9, 15, 56, 89
target, 12, 95
taxes, viii, 2
technical assistance, 28, 31, 34, 52, 53
technical support, 37
technologies, 39
technology, 20, 34, 39, 102
temperature, 25
territory, 38, 50, 65
testing, 28, 36, 71, 72
testing program, 28
Thailand, 32
thin market, 83
threats, 35, 39
time periods, 105
Title I, 9, 48, 89
Title II, 9, 48
Title IV, 9
Title V, 9, 31, 39, 48, 59
tobacco, xi, 57, 91, 94, 95
traceability requirements, viii, 2, 25
tracks, 41
trade, vii, 1, 8, 11, 28, 29, 30, 31, 32, 36, 37, 39, 60, 96
trade agreement, 30, 37
trade remedy, 28
trading partners, 30
training, 54, 82, 101
training programs, 54
traits, 100
transport, 99
transportation, 24, 53
tree nut, vii, 1, 2, 3, 4, 12, 14, 23, 24, 30, 38, 41, 47, 48, 54, 55, 57, 62, 90, 91, 95

U

U.S. Department of Agriculture, vii, ix, 1, 3, 68, 69
U.S. Geological Survey, 102
U.S. Treasury, 56
underwriting, 83, 90, 108, 109
uniform, 49, 75
unit cost, 82
United, v, 3, 23, 24, 26, 28, 30, 32, 33, 36, 37, 48, 53, 58, 62, 63, 93, 115
United Nations, 62
United States, v, 3, 23, 24, 26, 28, 30, 32, 33, 36, 37, 48, 53, 63, 93, 115
universities, 31, 34, 35, 38, 61, 103
Uruguay, 60
Uruguay Round, 60

V

varieties, 14, 55, 81, 83, 84
vegetables, viii, ix, 2, 3, 4, 8, 11, 12, 13, 14, 16, 20, 21, 22, 23, 25, 26, 28, 30, 36, 41, 42, 43, 45, 46, 47, 48, 51, 52, 54, 55, 57, 58, 62, 63, 64, 67, 69, 90, 91, 95
vegetation, 102
viruses, 24

W

wages, 54
waiver, 84
Washington, 3, 13, 17, 21, 29, 48, 52, 61, 78, 91, 92, 100, 110, 115
water, 24, 25
weakness, 72, 97
websites, 63
White Paper, 61
whole-farm insurance, ix, 68, 77, 84
wholesale, 20, 75, 108
WIA, 66
WIC, 20, 46, 47, 63, 64
wildlife, 32
windows, 105
Wisconsin, 52, 56, 65, 115
Women, Infants, and Children, 46
wool, 57
workers, viii, 2, 53, 54
workforce, 53
Workforce Investment Act, 66
working conditions, 54
workplace, 54
World Trade Organization, 37
worldwide, 39
WTO, 37, 60

Y

yield, viii, ix, xi, 67, 68, 69, 72, 73, 74, 75, 80, 81, 85, 86, 88, 91, 94, 103
Yugoslavia, 32